BIOTECHNOLOGY IN AGRICULTURE, INDUSTRY AND MEDICINE

# ION BEAM BIOENGINEERING RESEARCH

# BIOTECHNOLOGY IN AGRICULTURE, INDUSTRY AND MEDICINE

Additional books in this series can be found on Nova's website under the Series tab.

BIOTECHNOLOGY IN AGRICULTURE, INDUSTRY AND MEDICINE

# ION BEAM BIOENGINEERING RESEARCH

## S. ANUNTALABHOCHAI
## L.D. YU
## AND
## T. VILAITHONG

Nova Science Publishers, Inc.
New York

Copyright © 2011 by Nova Science Publishers, Inc.

**All rights reserved.** No part of this book may be reproduced, stored in a retrieval system or transmitted in any form or by any means: electronic, electrostatic, magnetic, tape, mechanical photocopying, recording or otherwise without the written permission of the Publisher.

For permission to use material from this book please contact us:
Telephone 631-231-7269; Fax 631-231-8175
Web Site: http://www.novapublishers.com

## NOTICE TO THE READER

The Publisher has taken reasonable care in the preparation of this book, but makes no expressed or implied warranty of any kind and assumes no responsibility for any errors or omissions. No liability is assumed for incidental or consequential damages in connection with or arising out of information contained in this book. The Publisher shall not be liable for any special, consequential, or exemplary damages resulting, in whole or in part, from the readers' use of, or reliance upon, this material. Any parts of this book based on government reports are so indicated and copyright is claimed for those parts to the extent applicable to compilations of such works.

Independent verification should be sought for any data, advice or recommendations contained in this book. In addition, no responsibility is assumed by the publisher for any injury and/or damage to persons or property arising from any methods, products, instructions, ideas or otherwise contained in this publication.

This publication is designed to provide accurate and authoritative information with regard to the subject matter covered herein. It is sold with the clear understanding that the Publisher is not engaged in rendering legal or any other professional services. If legal or any other expert assistance is required, the services of a competent person should be sought. FROM A DECLARATION OF PARTICIPANTS JOINTLY ADOPTED BY A COMMITTEE OF THE AMERICAN BAR ASSOCIATION AND A COMMITTEE OF PUBLISHERS.

Additional color graphics may be available in the e-book version of this book.

**LIBRARY OF CONGRESS CATALOGING-IN-PUBLICATION DATA**

Anuntalabhochai, S.
  Ion beam bioengineering research / authors, S. Anuntalabhochai, L.D. Yu, and T. Vilaithong.
       p. ; cm.
  Includes bibliographical references and index.
  ISBN 978-1-61209-529-5 (softcover)
  1. Ion bombardment. 2. Bioengineering. I. Yu, L. D. II. Vilaithong, Thiraphat. III. Title.
  [DNLM: 1. Ions. 2. Bioengineering--methods. QC 702]
  R857.I56A58 2011
  610.28--dc22
                                            2010051736

*Published by Nova Science Publishers, Inc. † New York*

# CONTENTS

| | | |
|---|---|---|
| **Preface** | | vii |
| **Author's Contact Information** | | ix |
| **Chapter 1** | Introduction | 1 |
| **Chapter 2** | Special Features | 5 |
| **Chapter 3** | Ion-Beam-Induced DNA Transfer and Ion Interaction with Cell Envelope | 13 |
| **Chapter 4** | Ion Beam Induced Mutation | 31 |
| **Chapter 5** | Summary | 67 |
| **Acknowledgements** | | 69 |
| **References** | | 71 |
| **Index** | | 75 |

# PREFACE

This review summarizes recent progress of a newly developed novel bioengineering technology, namely low-energy ion beam bioengineering (IBB), achieved at Chiang Mai University, Thailand. Low-energy IBB has demonstrated powerful impacts and highly potential applications on biology, agriculture, horticulture and life science owing to multiple-factor interaction between energetic ions and biological organisms as well as low cost and convenience in operation. Since late 1990s, IBB research programs have been vigorously carried out at Chiang Mai University. A group of scientists consisting of physicists, biologists, horticulturists, agriculturists, chemists, and medical scientists have developed special IBB ion beam lines and relevant techniques to use low-energy ion beam bombardment of biological organisms to induce mutation breeding and gene transfer. Besides the IBB applications, research interests are also focused onto basic mechanisms on ion interaction with DNA and biological cells to reveal physics and biology involved in the ion beam inductions of mutation and gene transfer. The results have been applied to serve developments of local agriculture and horticulture and promote national scientific research qualities. The contents include introductions to low-energy IBB techniques and facilities, and chapters on ion beam induced gene transfer, ion interaction with the cell envelope, ion beam induced mutation, and ion interaction with living cells and DNA.

# AUTHOR'S CONTACT INFORMATION

S. Anuntalabhochai[a,$], L.D. Yu[b,c,$,*], T. Vilaithong[c]
[a] Molecular Biology Laboratory, Department of Biology, Faculty of Science,
Chiang Mai University, Chiang Mai 50200, Thailand
[b] Plasma and Beam Physics Research Facility,
Department of Physics and Materials Science,
Faculty of Science, Chiang Mai University, Chiang Mai 50200, Thailand
[c] Thailand Center of Excellence in Physics, Commission on Higher Education,
328 Si Ayutthaya Road, Bangkok 10400, Thailand

---

[$] Contributed equally.
[*] Corresponding author: Tel.: +66 53 943379, Fax: +66 53 222776,
Email: yuld@fnrf.science.cmu.ac.th

*Chapter 1*

# INTRODUCTION

Plant seeds were carried by spacecrafts and sent to the space for mutation purpose because high-energy cosmic particles might irradiate and penetrate the seeds to induce mutation. However, scientists have found that they can achieve the same effect on the earth ground using low-energy particle bombardment of the seeds in a much cheaper, easier and more effective way. That is the initiation of the low-energy ion beam biotechnology. Low-energy ion beam biotechnology is such a technique that uses energetic ion beams (a few tens of kilo electron volts – keV – in energy being enough), generated and transported by an ion accelerator, to bombard biological organisms in vacuum to induce mutation breeding and gene transfer for increasing applications of the biological substance as well as to detect and analyze characteristics of biological organisms. This is a highly interdisciplinary field of physics, biology, agriculture and medical science. It is an extension of ion beam modification of conventional solid materials and an extension of high-energy radiobiology.

Ion beam biotechnology was invented by Prof. Yu Zengliang, Chinese Academy of Sciences, China, in late 1980s. Motivated and encouraged by the successful development of ion implantation modification of common solid materials, Prof. Yu made attempts to apply low-energy ion implantation techniques to the improvement of agricultural crop varieties [1,2]. Biological effects of ion implantation on rice were discovered. Since then ion implantation as a new tool for genetic modification has been applied to the breeding of crops and microbes. In 1988, Prof. Yu discovered the etching effect of ion beams on cells [3] and proposed the idea of ion beam processing of cells for gene transfer. Through the independent research of his Ph.D.

students, new varieties of gene-transferred rice were developed. Therefore, a new and highly interdisciplinary field of science, ion beam biotechnology or bioengineering (IBB), was completely established. The technology has been rapidly and extensively developed in China, where more than ten research laboratories nationwide are carrying out studies in this field and significant achievements have been attained [4]. In the beginning, except in China, IBB was studied only in a few laboratories in Japan [5] and USA [6], where the scientists worked based on high-energy charged particle radiation effects on biological cells, but not yet recognized by most of the scientists in the world. However, recently IBB has been recognized explosively and developed by many laboratories worldwide. For quite some time, heavy ions have been used for tumor therapy and micro-ion-beam probe technique has been developed for analysis of materials. Because of sound scientific and technological bases and traditions in Europe, recently European scientists such as in England [7] and Germany [8] have expanded and turned their research interests in the fields of heavy ion therapy and microbeam probe to a more comprehensive research field. In Thailand, at Chiang Mai University (CMU) we have known for quite some time that Prof. Yu Zengliang was the founder of ion beam biotechnology, and that he has carried out research in this field since the late 1980s. Stimulated by Prof. Yu's achievements, we organized and launched research in this area for serving economic development in Thailand. Only in a few short years, we have made great progress [9,10]. In the laboratory of Plasma and Beam Physics Research Facility (formerly Fast Neutron Research Facility), a specialized bioengineering ion beam line has been installed. Research on both fundamental interactions between ions and biological organisms and applications to agricultural and horticultural crop mutation breeding and gene transfer has been carried out by a collaborative group consisting of physicists, biologists and agriculturists. A number of important results have been achieved and published. We have been internationally leading in some areas of this field such as gene transfer in bacteria and yeast and ion interaction with the cell envelopes. For now, IBB has been one of the most ambitious research programs at Chiang Mai University. A chronicle of the main events of the IBB research is shown in Table 1. IBB has recently been rapidly expanded to the world [11]. There are nowadays three schools of the IBB scientists worldwide:

1. Represented by the laboratories in China, Japan and Thailand, the Asian school focuses research on ion beam induction of mutation and gene transfer.

2. Represented by the radiological research center at Columbia University and scientists at NASA, the American school focuses on radiation risk assessments of life and life science.
3. Represented by the Ion Beam Center at University of Surrey and laboratories in Germany, the European school focuses on ion beam analysis of biological organism materials and ion beam medical therapy.

In other continents, scientists in the countries such as South Africa, Brazil, Australia, etc. are also actively starting to be involved in this research. In this article we review achievements attained by the Thai scientists and also introduce some bases of IBB.

**Table 1. Chronicle of some important events of IBB research**

| Year | Event | Comment |
|---|---|---|
| 1988 – 1989 | • Wang Xuedong, Wu Yuejin, Yu Zengliang, et al., *Anhui Agricultural Science* (in Chinese), China, 1988.<br>• Wu Yuejin, Yu Zengliang, et al., *Anhui Agricultural Science* (in Chinese), China, 1989. | The first publications of IBB research, particularly on ion-beam-induced mutation breeding |
| 1989 | Yu Zengliang, et al., "Preliminary Study on Mechanisms of Ion Beam Induced Mutation Breeding", *Anhui Agricultural Science* (in Chinese), China, 1(1989)12-16. | The first proposition on the four-factors of ion beam induced biological effects |
| 1989 | R. Tanaka et al., "JAERI AVF Cyclotron for Research of Advanced Radiation Technology", *Proc. 12$^{th}$ Intern. Conf. On Cyclotrons and Their Applications*, Berlin, Germany, 1989, p.566. | The first work on IBB launched in Japan |
| 1991 | The 1$^{st}$ *National Symposium on Ion Implantation Induced Biological Effects*, Hefei, China, February 1991. | The first IBB-topical local symposium |
| 1993 – 1994 | • Wu Yuejin, Yu Zengliang, et al., *Proceedings of the 2$^{nd}$ National Symposium on Ion Implantation Induced Biological Effects*, Hefei, China, September 8-11, 1993, p.80.<br>• Yu Zengliang, et al., "Preliminary Study on Ion Beam Etching of Biological Samples", *Journal of Anhui Agriculture University*, 21(3)(1994)260-264. | The first studies on ion beam effect on etching of the cell wall, which is the theoretical basis of ion beam induction of gene transfer |
| 1993 | Yu Zengliang, et al., "Transferring Gus Gene into Rice Cells by Low Energy Ion Beam", *Nuclear Instruments and Methods*, B80/81(1993)1328-1331. | The first international report on ion-beam-induced gene transfer in plant cells |

## Table 1. (Continued).

| Year | Event | Comment |
|---|---|---|
| 1997 | T.K. Hei, L.J. Wu, S.X. Liu, et al., "Mutagenic Effects of a Single and an Exact Number of α Particles in Mammalian Cells", *Proc. Natl. Acad. Sci. USA*, 94(1997)3765–3770. | The first development and application of single-ion-beam technique for study of biological cells |
| 1998 | Yu Zengliang, "*Introduction to Ion Beam Biotechnology*" (in Chinese), Anhui Science and Technology Press, Hefei, 1998. | The first monographic book on IBB published |
| 1999 | T Vilaithong, L D Yu, C Alisi, B Phanchaisri, P Apavatjrut, S Anuntalabhochai, "A Study of Low-Energy Ion Beam Effects on Outer Plant Cell Structure for Exogenous Macromolecule Transferring", *Surface and Coatings Technology*, 128-129(2000)133-138. | The first introduction of Thai researchers' IBB work to the international community |
| 2001 | S Anuntalabhochai, R. Chandej, B Phanchaisri, L.D. Yu, T. Vilaithong, I.G. Brown, "Ion-Beam-Induced Deoxyribose Nucleic Acid Transfer", *Applied Physics Letters*, Vol.78, No.16(April 16, 2001)2393-2395. | The first international report on ion-beam-induced gene transfer in bacteria |
| 2002 | *The First International Symposium on Ion Beams: Biological Effects and Molecular Mechanisms*, July 30 – August 2, 2002, Wulumuqi, China, | The first international symposium on IBB |
| 2004 | The Network on ion-beam bio-medical applications launched in England. | The first large-scale research projects on IBB launched in Europe |
| 2006 | Yu Liangdeng, Thiraphat Vilaithong, Ian Brown, translated Yu Zengliang's "*Introduction to Ion Beam Biotechnology*" (English Edition), Springer Science & Business Media, New York, 2006. | The first classic textbook on IBB internationally published |
| 2007 | • P.H. Mao, et al., "Total DNA of Chinese ephedra transferred to yeast by ion beam", Chinese patent No.200610011402.<br>• S. Anuntalabhochai, et al., "Ion-beam-induced gene transfer in *Saccharomyces cerevisiae*", *The 15th International Conference on Surface Modification of Materials by Ion Beams*, Mumbai, India, September 30 - October 5, 2007 | The first evidence of gene transfer in yeast |
| 2009 | Project THA 5049 of IAEA (International Atomic Energy Agency) entitled "Establishing an Ion Beam Biotechnology Center in Thailand to develop improved crops for agriculture and horticulture" started. | IAEA was the first time involved in supporting IBB for developing countries |

*Chapter 2*

# SPECIAL FEATURES

## 2.1. COMPLEXITIES

Ion beam interaction with biological living organisms is so different from that with normal solid materials. The latter has been studied for many decades by physicists and thus related physics (only) is clear and simpler compared with the former, which is considerable complicated with many unknowns remaining. Some of the complexities are listed below

- Biological organisms are living, and ion beam treatment should not kill them.
- Fresh cells contain a large amount of water, which essentially evaporates in vacuum, and the evaporation causes differences in the target status from that in normal atmosphere.
- Biological material structures are highly porous and inhomogeneous, and ions penetrate and sputter abnormally more than for normal condensed materials.
- The functioning structures of organisms are very complicated and different ion-beam treated locations have different responses, and hence in order to get a certain response, ion beam should be precisely controlled to target the location.
- Biological organisms are extremely sensitive to ion irradiation, actively respond to the irradiation and thus highly produce secondary effects, which can greatly affect consequences of ion beam bombardment.

- Organisms have recovery ability, and ion beam radiation damage may be repaired and thus ion beam effects may be significantly modified and even eliminated in a certain time period.
- Different parts of an organism may have communications and an ion-beam treated location may produce unexpected effects.

## 2.2. ION BEAM FACILITIES

### 2.2.1. Requirements

Following requirements for ion beam facilities to perform IBB research are recommended:

- Large, homogeneous and high-current beam for mutation. Because ion beam induced mutation is basically a random process, in which a large number of samples should be treated to produce some desired mutations.
- Precise beam in energy, fluence and size for gene transfer. Because ion beam should process precisely only the cell envelope, in which radiation-damage-induced pathways can be created for exogenous gene to pass through.
- Micro/single-ion beam for life science study. Because life science study requires the ion(s) to only bombard certain locations at a single cell.
- Vertical beam line. Because horizontal holding of biological samples of irregular size and shape is convenient.
- Sterile environment. Because treatments of biological organisms should avoid contamination from unexpected organisms.
- High pumping power. Because living organisms can survive in low pressure conditions only for short time.

To meet these requirements, one may design and construct specially IBB-purposed ion beam facilities, or modify conventional ion implanters for IBB applications.

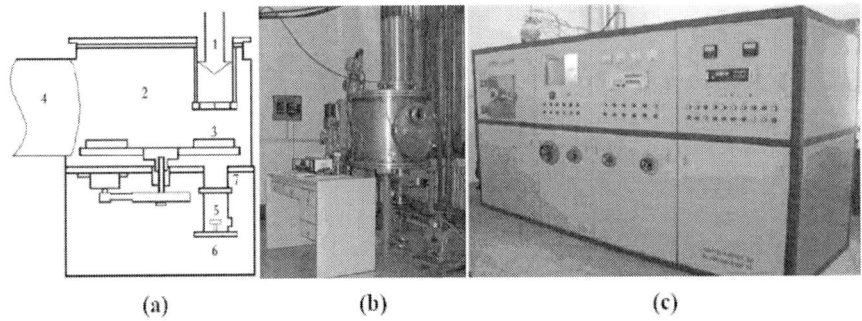

(a)  (b)  (c)

Figure 1. An example of IBB ion implanter. (a) Schematic diagram of the treatment chamber of the IBB facility at the Ion Beam Bioengineering Laboratory, Institute of Plasma Physics, Chinese Academy of Sciences, at Hefei (Reproduced from Fig. 2.8, page 20, Chapter 2 of [11], Copyright (2006), with the kind permission of Springer Science and Business Media). (1) Ion beam, (2) Big target chamber, (3) Rotating plate, (4) Vacuum pump, (5) Small target chamber, (6) Sterile chamber, and (7) Valve. (b) Photograph of the equipment. (c) The control board of the ion implanter.

An example of typical ion implanters for IBB purposes is shown in Figure 1. The ion beam extracted/accelerated from a broad-beam ion source is incident vertically into the chamber. The chamber is 90 cm in diameter and 1.0 m in height. A rotating plate of 80 cm in diameter is on the bottom of the chamber. On the plate there are six 20-cm-diameter concave holes, one of which is completely through. Six 20-cm-diameter sample dishes can be fixed in the concave holes. On the sample dishes there are holes with different diameters for holding various bio-samples. Samples of the same variety, such as rice seeds, are placed in the sample dishes, which are then fixed onto the rotatable big plate. The angular position of the plate is adjustable to align one of the dishes with the incident ion beam. After implantation of the first dish, other five sample dishes can automatically be aligned in sequence to the ion beam by computer control, which can also set different implantation parameters. Thus six different sample groups with different implantation parameters can be obtained in a single equipment operation cycle. Ion implantation of microbes or plant cells is normally carried out in a small target chamber under the main chamber connected by a gate valve. The samples are installed under sterile conditions. The hole on the rotating plate is aligned to the entrance of the small chamber and the valve is then opened to allow ion implantation.

## 2.2.2. IBB Ion Beam Facilities at CMU

There are two IBB ion beam facilities available at CMU.

### *1. Bioengineering Ion Beam Line*

This beam line possesses special features [12]: vertical main beam line, low-energy (30 keV) ion beams, double swerves of the beam, a fast pumped target chamber, and an *in-situ* atomic force microscope (AFM) system [13] chamber, as shown in Figure 2. The whole beam line is situated in a bioclean environment, occupying two stories. The quality of the ion beam has been studied. It has proved that this beam line has significantly contributed to our research work on low-energy ion beam biotechnology.

The ion source (Danfysik model 910) is able to produce various ion species from gas, solid, metal and nonmetal source materials. The beam current can be up to hundreds of microampere. The maximum accelerating voltage is 30 kV. The $90°$ mass-analyzing magnet is able to select any ion species in principle. The magnet is supported and fixed in vertical by a four-post steel framework capable of being adjusted (shifted) in position both horizontally and vertically. The beam line after the mass analyzing magnet is vertical in order to accommodate convenient horizontal holding of biosamples which are normally very difficult to hold in vertical because of their irregular and various shape and size. The Faraday cup and beam profile monitor are used to measure the beam current and density distribution along the beam line. The double-magnet beam steering system consisting of two beam sweeping magnets bends the ion beam twice in a small angle. The second magnet also acts as a beam sweeper. Neutral particles may not be a serious problem to ion beam modification of normal solid materials, but indeed a great disturbance to sensitive biosamples. Conventionally avoiding the neutral particles is simple, just using a small beam bending device, either electrostatic or magnetic. However, if only one magnet is used, although the neutral particles are avoided, the beam in our facility no longer vertically enters the target chambers, particularly the *in-situ* AFM chamber, which is designed to definitely require the beam to vertically enter.

The target chamber is deliberately made small, about 20 cm in diameter and 20 cm in height for fast pumping to reduce the risk for biosamples staying in vacuum too long. In routine operations, the target chamber can be pumped from the atmosphere to the working pressure of an order of $10^{-3}$ Pa within 10 minutes. Inside the chamber there is a horizontally moveable sample holder, controlled by a stepping motor, which holds a standard biological petri dish where the bio-sample is placed.

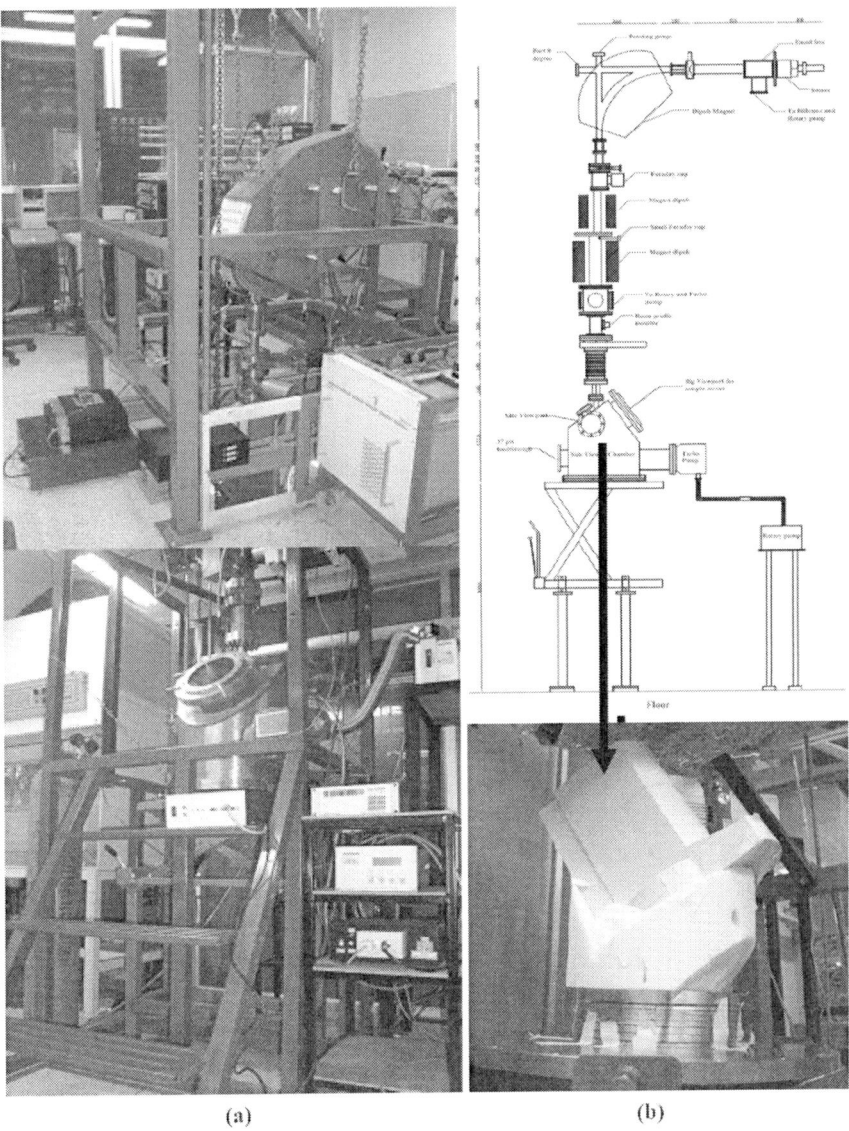

Figure 2. The bioengineering ion beam line at CMU. (a) Photograph. The upper one is the part upstairs and the lower one is the part downstairs. (b) Schematic diagram with a photo of the *in-situ* AFM sitting inside the chamber.

Cooperation of the sample holder translation and beam scanning makes a bombarding area of 5 cm × 5 cm in maximum. An automatically controlled halogen-lamp heating system is installed to maintain the bio-sample temperature for experiments on living organisms. Below the target chamber is the *in-situ* AFM system chamber for real-time surface observation of biosamples. In brief, a commercial AFM setup is placed inside the chamber in a certain tilting angle to allow the well-collimated ion beam to bombard the sample without hitting any parts of the microscope and with a damping system to avoid external vibration disturbances. Both tilting angle and damping system were carefully designed in geometry and mechanics [14]. There is a convenient control board to control the operational parameters of the ion source, gas inlet, mass analysis, and beam bending and scanning, the measurement of the beam current, and the manipulation of the target holder. The bioengineering ion beam line facility is entirely housed in an ion-beam-bioengineering compartment.

## *2. High-Current Broad-Beam Ion Implanter*

The ion implanter is versatile non-mass-analyzing ion implantation equipment [15], as shown in Figure 3. The facility was originally developed for industrial applications and later modified for bioengineering purpose as well. The ion source is a duoplasmatron source which is able to produce high ion-beam currents and broad beam of nitrogen ions. The maximum accelerating voltage is 150 kV. As the ion implanter does not have a device such as a magnet for ion mass analysis, the ion beam consists of both atomic and molecular nitrogen ion species. There are two target chambers. The small chamber is used for material treatment that requires short time as the chamber can be pumped down fast. The big main chamber is used for normal treatment of samples which may be in big size or complex shape. Inside the main chamber a special sample holder stage is installed for complicated treatment operation. The stage is able to rotate in two dimensions and translate. A specially designed bio-sample holder can be placed on the stage. The holder is able to hold samples on its four sides to carry out ion beam bombardment simultaneously. It has an ability of water cooling to prevent the target from overheating. For ion beam mutation, biological samples such as rice seeds are placed in standard Petri dishes using sticker tape for fixing, or in special seed holders (Figure 3). On each seed holder, made from copper with 10 cm in diameter, hundreds of small holes are made and each hole holds a seed. The seed holders are then installed onto the sides of the target holder for bombardment. The ion beam bombardment of biological samples is operated in a pulse mode normally with beam exposure and cutoff times of minutes respectively to avoid overheating the samples.

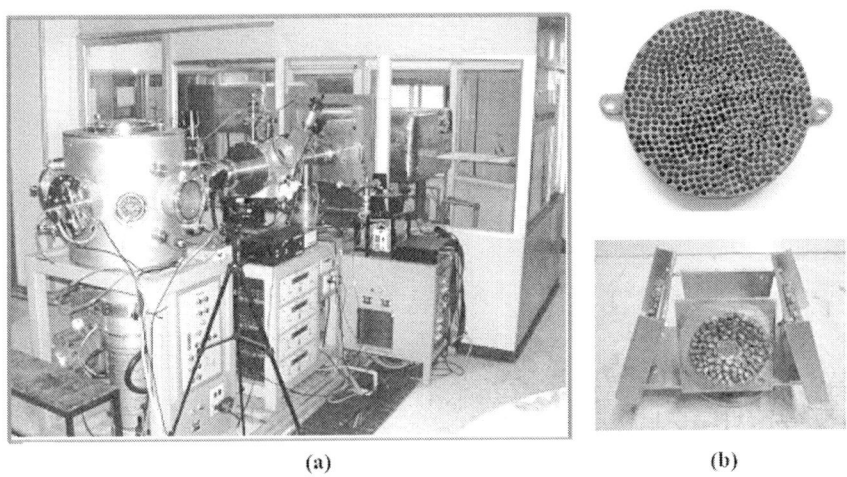

Figure 3. Photographs of the versatile ion implanter (a) and the equipped special bio-sample holders (b) for IBB at CMU. In (b), the upper one is the copper-made holder holding rice seeds, and the lower one is the four-side holder holding corn seeds.

*Chapter 3*

# ION-BEAM-INDUCED DNA TRANSFER AND ION INTERACTION WITH CELL ENVELOPE

## 3.1. INTRODUCTION

Using ion beam bombardment is a physical method to induce DNA transfer in cells. In the method, the ion beam parameters such as ion species, energy and fluence are precisely controlled so that the ions can only bombard the cell envelope to create radiation-damage-induced special structures which are expected to act pathways for exogenous macromolecules to pass through. The range and extent of the radiation damage for the target material of the cell envelope depend on the ion beam parameters as mentioned above. With given energy and fluence, lighter ion species penetrate deeper in the target material but create lighter damage, whereas heavier ions have shorter ranges but produce heavier damage. With given ion species and fluence, the higher energy the ions have, the greater penetration depth and radiation damage the ions can lead to. With given ion species and energy, the higher fluence the ion beam has, the more the surface sputtering which results in deeper penetration of the ions and the more the radiation damage. Too short penetration depth and light damage in the material may not create sufficiently deep and large pathways, while too long ion range and heavy damage may kill the cell. Therefore, the ion beam parameters should be carefully designed for treating different biological cell species which have different cell envelope thicknesses. The post biological treatment for DNA transfer is carried out usually immediately after ion beam bombardment normally in standard biological procedures.

## 3.2. EXPERIMENTS

The experiments for studies on ion beam induction of DNA transfer and related mechanisms were operated mainly using the bioengineering-specialized ion beam line. Ion species of N, Cl, Ar, Xe, Mg, Al, Ti, Fe, Ni, Cu, Au, etc. were used to bombard biological target materials of bacterial cells of *Escherichia coli* (*E. coli*) (strain DH5α) for DNA transfer and cells of onion skin, corn embryo and *Curcuma* embryo for study of mechanisms. The ion energies were ranged from 20 keV to 30 keV and the fluences normally ranged from $10^{15} - 10^{16}$ ions/cm$^2$. Post-bombardment biological treatments followed relevant standards.

## 3.3. RESULTS

### 3.3.1. Vacuum Effect

Vacuum could cause effects on survival and modification of the living materials because of severe water loss [16,17], low temperature, and deformation of the materials, and thus effect if any due to vacuum should be first screened. It was observed that in the vacuum condition such as the pressure of about $10^{-4} - 10^{-3}$ Pa, cells considerably shrank (Figure 4) [17] due to severe water loss and the temperature was around or lower than 0°C. However, for an appropriate vacuum exposure time duration, cells survived and lives were maintained in normal (Figure 5) [17]. Post biological treatment could not transfer exogenous macromolecules into the cells exposed to vacuum only (Figure 6) [17].

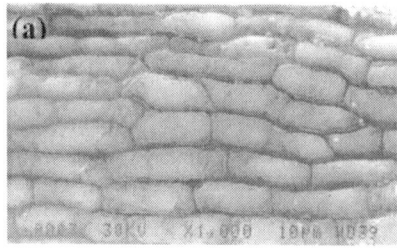 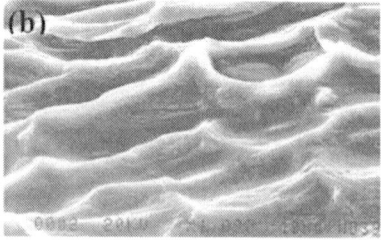

Figure 4. SEM images of living onion cells (Reprinted from [17], Copyright (2007), with permission from Elsevier). (a) Fresh control, and (b) under vacuum.

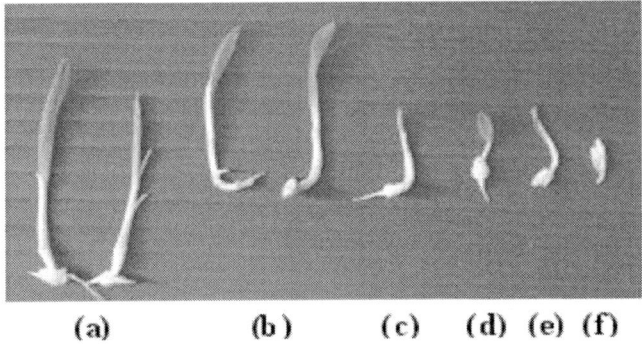

(a)    (b)    (c)   (d) (e) (f)

Figure 5. Germination of naked corn embryos after 15-keV N-ion bombardment (Reprinted from [17], Copyright (2007), with permission from Elsevier). (a) Fresh control, (b) vacuum control, (c) – (f) bombarded with fluences of $5 \times 10^{14}$, $1 \times 10^{15}$, $2 \times 10^{15}$, and $1 \times 10^{16}$ ions/cm$^2$, respectively. The vacuum control is seen to have similar germination with the fresh control, whereas the ion-bombarded seeds germinated badly and at the highest fluence the seed seemed almost dead.

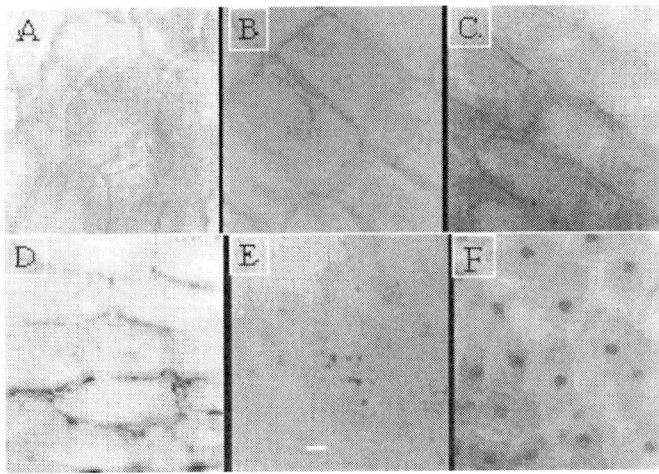

Figure 6. TB (Trypan-blue)-staining of corn embryo cells bombarded by 30-keV Ar ion beam (Reprinted from [17], Copyright (2007), with permission from Elsevier). A: fresh control, B: vacuum control, C – F: ion bombarded with fluences of $5 \times 10^{14}$, $1 \times 10^{15}$, $2 \times 10^{15}$, and $4 \times 10^{15}$ ions/cm$^2$, respectively. The TB-stained areas indicate the locations where the dye molecules enter. For the vacuum control, the dye does not enter the cells, but for the ion-bombarded cells the dye enters either the cell envelope or inner, depending the fluence. TB only stains dead parts of the cell, and staining of the cell nuclei indicates death of the cells. The darker areas are TB-stained. The white bar in the figure indicates 5 μm.

## 3.3.2. Survival

In order to obtain knowledge on the critical ion beam conditions that still maintain living cells alive for DNA transfer into the cells, for various ion species, ion energy and fluence were varied. For example, when Ar ion beam at 30 keV was used to bombard corn embryo cells, the critical fluence was found to be $2 \times 10^{15}$ ions/cm$^2$ [18,19] (Figure 5). Fluences higher than the critical fluence could cause death of the cells, whereas fluences too lower than the critical condition could not result in exogenous molecule transfer into the cell (Figure 6) [17].

## 3.3.3. DNA Transfer

Using the critical ion beam conditions, which could lead to the exogenous molecule transfer in the optimal efficiency, we succeeded in DNA transfer in bacterial cells of *E. coli* [20]. Figure 7 shows the introduction of pGEM-T easy and pGFP plasmids, containing *lacZ* and *GFP* genes respectively, into *E. coli* strain DH5α bombarded by Ar ions at an energy of 26 keV and a fluence of $2 \times 10^{15}$ ions/cm$^2$. The blue and green colonies obtained by incubation of the bacteria in medium supplemented with X-gal-substrate for *lacZ* gene's product and activation of *GFP* under UV, respectively, demonstrate that the DNA has been indeed transferred into the bacterial cells. The subsequently measured DNA molecular sizes further confirm that the transferred DNA is the original exogenous DNA. Figure 8 demonstrates the ion-beam-induced transfer of plasmid DNA pGEM2 that contains ampicillin resistance gene in bacteria *E. coli.* [21].

In another work, two marker genes, *GFP* and *lipoic acid synthetase,* were chosen to transfer into yeast (*Saccharomyces cerevisiae* strain W303C, *MAT*a *ura3-52 his3-Δ200 lys2-801*) [22]. The yeast cells were bombarded by nitrogen ions at energy of    keV with fluences of    ,   , and    $10^{16}$ ions/cm$^2$, and subsequently the bombarded yeast cells were incubated with both plasmids, pYGFP and pYlip plasmids, carrying *GFP* and *lipoic acid synthetase* genes separately. The expression of *GFP* gene in the yeast was observed by green yeast colonies under UV light (Figure 9a), while the expression of *lipoic acid synthetase* gene was analyzed using the sodium dodecyl sulfate polyacrylamide gel electrophoresis (SDS-PAGE) method. A gene's product at 34 kDa was detected only in the bombarded yeast with the pYlip (Figure 9b). The expression of both genes was induced by culturing the yeast cells in YPD (composed of yeast extract, peptone and glucose) media supplemented with galactose for 10 hrs. These evidences demonstrated that nitrogen ion bombardment assisted gene transfer in the yeast cells. This work has paved a road for promoting yeast bioengineering.

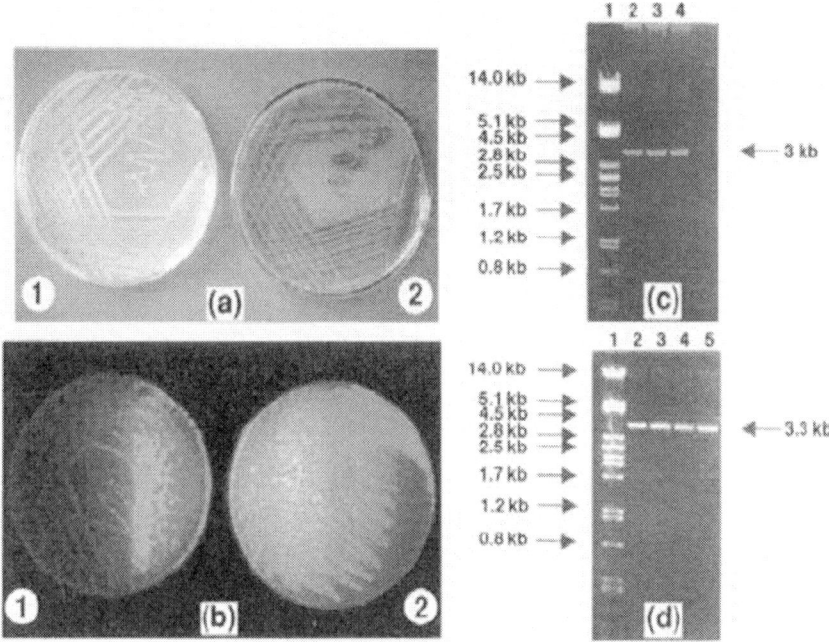

Figure 7. Expression of the ion-beam-induced gene transfer in the cells of bacteria *E. coli* and the corresponding molecular size measurement (Reprinted with permission from [20]. Copyright [2001], American Institute of Physics). Left [(a) and (b)]: gene expression of *lacZ* and *GFP* genes in Luria broth (LB) medium. Right [(c) and (d)]: molecular size measurement by electrophoresis. (a) and (c): expression and molecular size of transferred pGEM-T easy plasmid, respectively. (b) and (d): expression and molecular size of transferred pGFP plasmid, respectively. (1) Without gene, and (2) with gene transferred. In (c) and (d), lane 1 shows the standard molecular size markers, and other lanes show the DNA molecular sizes measured by various restriction enzymes. The measured molecular sizes are confirmed the same as those of the originally exogenous DNA.

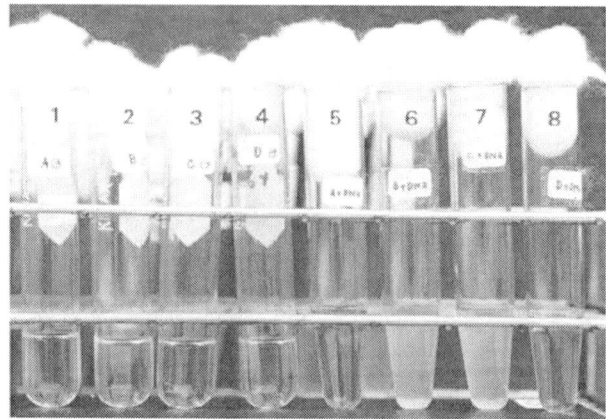

Figure 8. Demonstration of ion-beam-induced transfer of plasmid DNA pGEM2 that contains ampicillin resistance gene in bacteria *E. coli*. (Reprinted from [21], Copyright (2002), with permission from Elsevier). 26-keV Ar-ion-beam bombarded bacteria with fluences of 0.5, 1, 2, and 4 × $10^{15}$ ions/cm$^2$ (Tubes 1 and 5, 2 and 6, 3 and 7, and 4 and 8, respectively) were incubated in antibiotic medium. Tubes 1, 2, 3, 4: without DNA transfer. Tubes 5, 6, 7, 8: with DNA transferred. The turbidity indicates the growth of the bacteria.

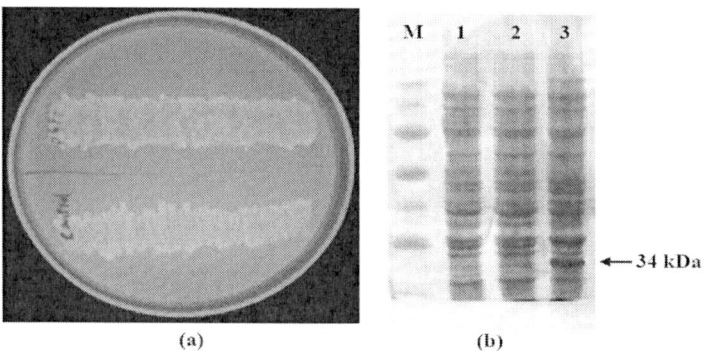

Figure 9. Demonstration of ion-beam-induced transfer of plasmid DNA, pGFP and pYlip containing *GFP* and *lipoic acid synthetase* marker genes, respectively, in yeast (*S. cerevisiae* strain W303C) bombarded by nitrogen ions at energy of    keV and fluences of    $10^{16}$ ions/cm$^2$. (a) A green colony of yeast indicates an expression of *GFP* gene in the yeast cells. (b) SDS-PAGE analysis of *lipoic acid synthetase* in the yeast transferred with ion beam (Reprinted from [22], Copyright (2009), with permission from Elsevier). (1) *S. cerevisiae* wild-type, (2) transformed but non-induced, (3) transformed and induced recombinant *S. cerevisiae*, and (M) molecular mass marker. The arrow indicates the band (M~34 kDa) corresponding to recombinant lipoic acid synthetase. The gel was stained with Coomassie blue to visualize the protein bands.

## 3.4. DISCUSSION

How low-energy ions could induce exogenous molecules to transfer into the cell is the question we have always been interested in. Can ion beam bombardment induced radiation damage in the cell envelope be deep and large enough to form pathways to allow the exogenous macromolecules to pass through? Or may other non-physical pathway mechanisms be involved? We have made intensive investigations and discovered some interesting phenomena.

### 3.4.1. Abnormally Great Ion Range and High Sputtering

Low-energy ions have short ranges in normal materials. The range of Ar ions of 25 keV in cellulose, which is the main plant cell wall material, with a chemical composition of $C_6H_{12}O_6$ and a mass density of 1 g/cm$^3$ is calculated to be about 50 nm, a negligible number of the ions can penetrate to the depth of 100 nm and a sputtering loss of the surface thickness is 6 nm according to PROFILE [23]. However, the experimentally observed range and sputtering loss of 25-keV Ar ions in the plant cell envelope are about 200 nm [24] and hundreds of nanometers, respectively (Figure 10) [25]. Therefore, in reality the ions have abnormally greater range and sputtering than theoretical estimates in the cell envelope materials. The factors of both great range and sputtering can cause ion penetration considerably deeper than those expected in the cell envelope. Hence, it is possible for ion beam to create radiation damage in an extensive region in the cell envelope material.

### 3.4.2. A Physical Model of the Cell Wall

In order to interpret the observed abnormally great ion range and sputtering, we set up a physical model of the plant cell wall [24] (Figure 11 [17,25]). According to the model, the linear thickness of the modeled cell wall material is 7/17 of the fresh cell wall, the mass density is 1 g/cm$^3$, the average number of bonds of each atom in a molecule is 2, and therefore, the stopping cross section is deduced to be about 1/2, the ion range about 2.5 times and the sputtering yield about 7 times those for a matrix with homo-distributed atoms of the same material, respectively [24]. With this model, the estimated ion depth distribution in the cell envelope is fairly close to the measured one, as shown in Figure 12 [17,26].

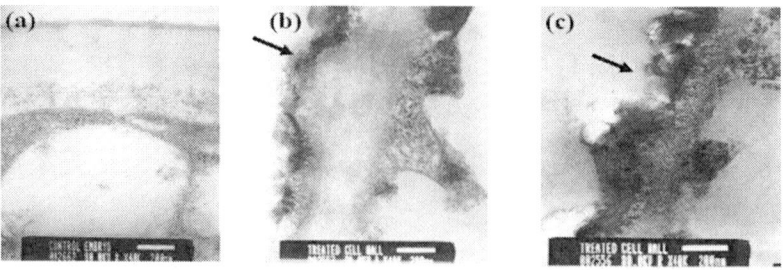

Figure 10. Transmission electron microscopy (TEM) images of cross sections of the Curcuma embryo cell envelopes (Reprinted from [17], Copyright (2007), with permission from Elsevier). (a) Control, (b) and (c) Ar-ion bombarded with fluences of $1 \times 10^{15}$ ions/cm$^2$ and $2 \times 10^{15}$ ions/cm$^2$, respectively. The arrows point the external surface of the cell envelope.

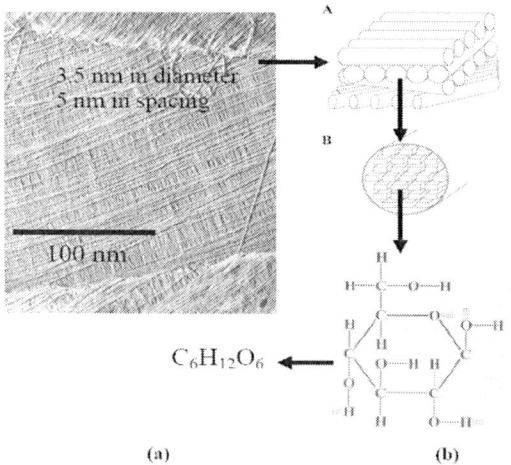

Figure 11. The physical model of the cell wall (Reprinted from [17], Copyright (2007), with permission from Elsevier). (a) The real cell wall is a discontinuous structure consisting of nanosized cellulose microfibrils, made from cellulose molecules in a chemical structure of $C_6H_{12}O_6$, in a size of 3.5 nm in diameter (for most of high plants) and cross-linked in a net style with about a 5-nm spacing in between [27,28]. (b) The physical model of the cell wall is therefore a compact solid, which has a linear thickness of 7/17 that of the real cell wall, consisting of 3.5-nm-diameter microfibrils arranged in parallel in a plane but oriented in random for different planes (A); each microfibril is composed of chains of cellulose molecules in the chemical structure of $C_6H_{12}O_6$ (B).

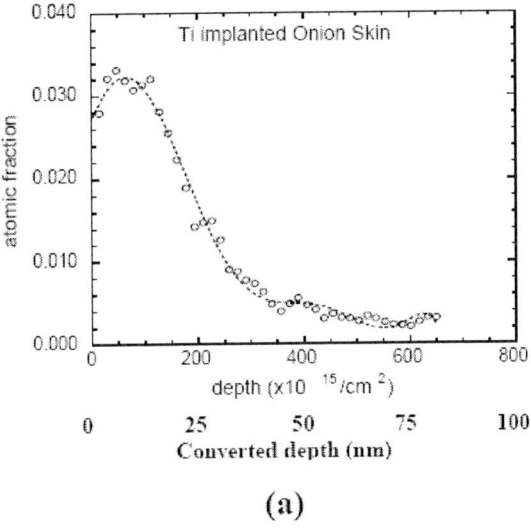

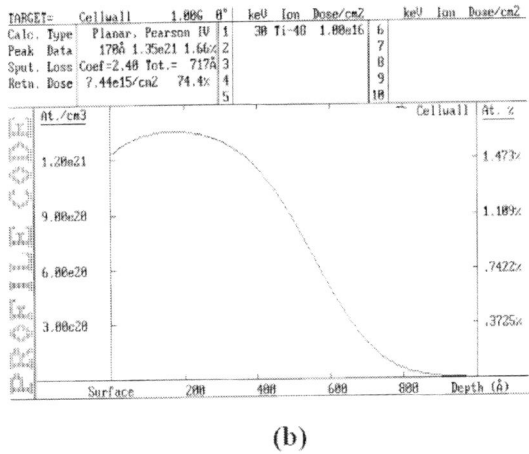

Figure 12. The implanted Ti-ion concentration distribution as a function of the depth in the onion cell envelope (Reprinted from [17], Copyright (2007), with permission from Elsevier). The ion implantation energy is 30 keV and the applied fluence is $1 \times 10^{16}$ ions/cm$^2$. (a) Measured using Rutherford backscattering spectrometry (RBS), and (b) calculated using PROFILE computer code based on the physical model.

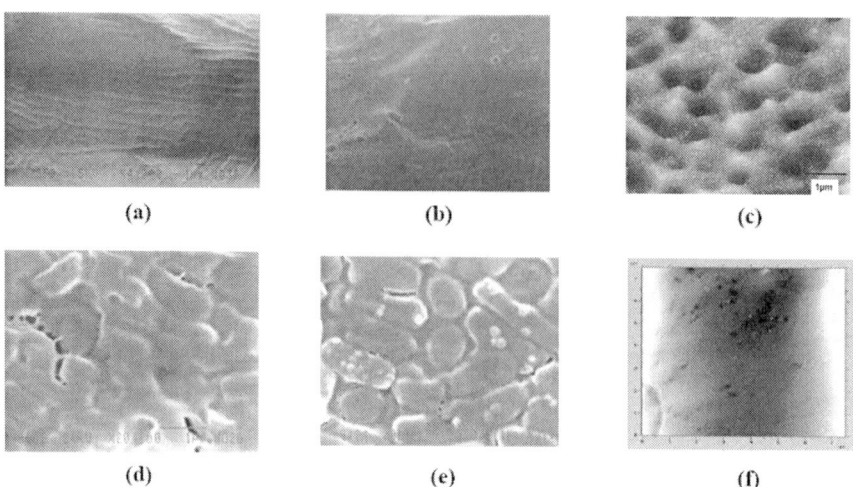

Figure 13. Low-energy ion bombardment induced formation of micro/nano-craters in the plant cell envelope. (a) SEM photograph of the surface of the unbombarded onion skin cells. (b) Ex-situ SEM photograph of the 30-keV Xe-ion bombarded onion skin cell. (c) Ex-situ AFM image of the 25-keV Ar-ion bombarded onion skin cell. (d) SEM photograph of the unbombarded bacterial *E. coli* cells. (e) Ex-situ SEM photograph of the 25-keV Ar-ion bombarded *E. coli* cells. (f) In-situ AFM image of the 25-keV Ar-ion bombarded onion skin cell. (Reprinted from [17,26,30], Copyrights (2007, 2005, 2007), with permission from Elsevier). Micro/nano-crater-like structures are seen at all of the ion-bombarded cell envelopes.

### 3.4.3. Micro/Nano-Craters

After ion bombardment of plant and bacterial cells, micro/nano-craters were observed either *ex-situ* or *in-situ* using scanning electron microscopy (SEM) and atomic force microscopy (AFM) [13,25,26,29,30], as shown in Figure 13. Molecular dynamics simulation of ion beam bombardment of the plant cell wall was also carried out for studying the formation of the crater structure (Figure 14), but preliminary results did not show any possibility from the ion direct interaction with the biological molecules. Some possible conclusions can be drawn from the experimental fact of the formation of the micro/nano-craters:

- The formation of the micro/nano-craters is a general phenomenon induced during ion bombardment, no matter what ion species, under

certain ion beam conditions such as ion beam energy and fluence range;
- The craters are inhomogeneously distributed, thus the crater formation is not a direct effect but maybe an indirect consequence of ion bombardment;
- The effect is not related to the material chemical composition and rapid water evaporation from the cell envelope;
- It is related to the special microstructure of the cell envelope;
- The craters may be pathways for exogenous DNA transfer into biological cells;
- Evidence for the craters to act as DNA transferring pathways is insufficient and to be further explored;
- If the craters are demonstrated to act indeed as the pathways, the driving force of the transferring remains unclear yet.

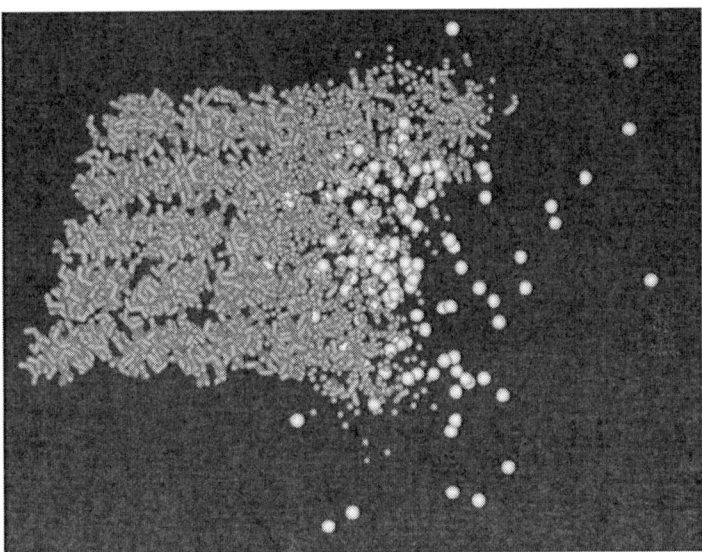

Figure 14. Molecular dynamics simulation result of the cellulose structure after $Fe^{2+}$ bombardment for 0.01024 ps (Reprinted from [17], Copyright (2007), with permission from Elsevier). The velocity scale is 1/10 the actual velocity (energy: 25 keV). The lighter particles are Fe ions and the darker particles are the atoms structuring the cellulose. It is seen that at the ion-beam spot bombarding the target material, the ions generally displace the atoms of the cellulose while the ions are penetrating.

## 3.4.4. Using Polymeric Membrane Mimics to Simulate Ion Bombardment of the Plant Cell Envelope and Study Relevant Mechanisms

Due to difficulties in getting useable monolayer of plant cell envelopes for basic mechanism study, based on the similarly chemical structure and compositions of both chitosan ($C_6H_{11}O_4N$) and cellulose with the plant cell wall in which cellulose ($C_6H_{12}O_6$) is the main composition, chitosan and cellulose membranes were used as simple and easy mimics of the plant cell envelope to simulate and characterize relevant behavior in order to separately investigate effects of ion interaction with plant cell envelope [31]. In the experiments, square samples of chitosan and cellulose membranes with nitrogen and argon ions at energy of 15-25 keV to fluences in order of $10^{15}$ ions/cm$^2$ which were the same as those used for ion-beam induced DNA transfer. The membrane surface morphology was investigated with atomic force microscopy (AFM) in the dynamic mode. The contact angles of the membranes were measured for their biocompatibility. The membranes were characterized by electrical membrane impedance spectroscopy using a two-chamber system [32] to study the membrane electric properties under a 10 mM KCl solution condition.

No matter the chitosan or cellulose membrane surface, although the former looks smoother than the latter, after ion bombardment the surface generally becomes rougher (Figure 15). The membrane surface roughness is increased as increasing the ion fluence but normally decreased as increasing the ion energy, as shown in Figure 15c. This change is more obvious for Ar-ion bombardment. From the results it is seen that membrane surface is more sensitive to argon ion bombardment. This demonstrates that Ar-ion bombardment has been more effective in induction of gene transfer. With ion bombardment, the surfaces of the chitosan and cellulose membranes behaved differently in the contact angle. The contact angle of the chitosan membrane decreased as increasing the ion fluence, whereas that of the cellulose membrane showed no regular patterns. The changes in the contact angle seem not related to the membrane surface roughness. As the cellulose is the main composition of the plant cell wall, the results indicate that the modification of the contact angle or hydrophilicity of the plant cell envelope surface is not a key factor for ion-beam induced gene transfer.

The membrane impedance spectroscopy provides the values of the impedance or conductance and capacitance of the membranes, as shown in Figure 16. It is seen that ion bombardment considerably decreases the impedance of both membranes, especially in the cases of N-ion bombardment and the cellulose membrane. The decrease in the impedance indicates a favor to transferring charged substance such as DNA. Ion bombardment can increase the capacitance of the membranes. This is understandable as the ion bombardment increases the membrane surface area by roughening and reduces the membrane thickness by sputtering. The increase of the capacitance indicates an increase in the driving force for transferring charged substance such as DNA molecules.

In order to demonstrate the consequence of ion bombardment effect on the membranes in DNA transfer, an *in vitro* experiment on plasmid DNA transfer through the cellulose membrane was conducted. The membrane was bombarded with N-ion beams at 25 keV and fluencies of 1, 2, 4 × $10^{15}$ ions/cm$^2$ and followed by dropping plasmid DNA, pGFP, onto the top side of the membrane under which was a small 200-ml PCR tube fully containing deionized water to collect the transferred DNA (Figure 17). After 30 minutes the solution in the PCR tube was characterized by a DNA spectrometer for the DNA concentration and gel electrophoresis for confirming the transferred DNA. The measured DNA concentration in the solution in the PCR tube, as shown in Table 2, is significantly higher for the ion-bombarded membrane than the unbombarded control and the highest at the fluence of 2 × $10^{15}$ ions/cm$^2$. This result clearly shows that DNA transfer is absolutely favored by ion bombardment. Interestingly note that the impedances of the membranes are the lowest at the N-ion fluence of 2 × $10^{15}$ ions/cm$^2$. This coincidence further indicates the role of the impedance played in the DNA transfer. The electrophoresis analysis result shown in Figure 17b confirms the plasmid DNA that has been transferred through the membrane is the same as that at the top of membrane as they have the same band positions. This fact demonstrates that the DNA detected at one side of the membrane is indeed the original DNA at the other side of the membrane but not contaminant.

These results revealed that the increase in the surface roughness which provides more surface contact area, decrease in the membrane impedance which can accelerate the DNA transfer, and increase in the membrane capacitance which raises the driving force to transfer DNA are involved in the mechanisms for ion-beam-induced gene transfer, but the contact angle which is an indicator of hydrophilicity or hydrophobicity of the materials surface is not.

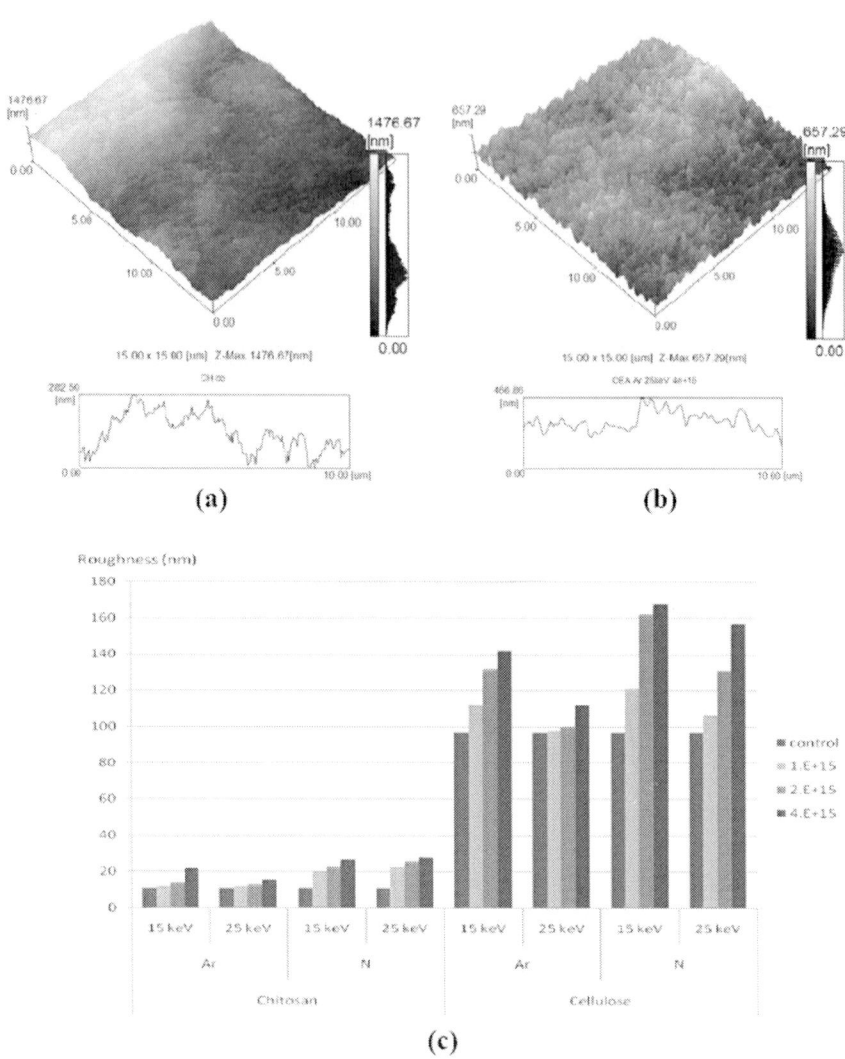

Figure 15. Membrane surface roughness modified by ion bombardment. (a) AFM image of the cellulose control surfaces. (b) AFM image of the cellulose surface bombarded with 25-keV Ar ions to a fluence of $4 \times 10^{15}$ ions/cm$^2$. (c) Summary of the membrane surface roughness changes (Reprinted from [30], Copyright (2009), with permission from Elsevier).

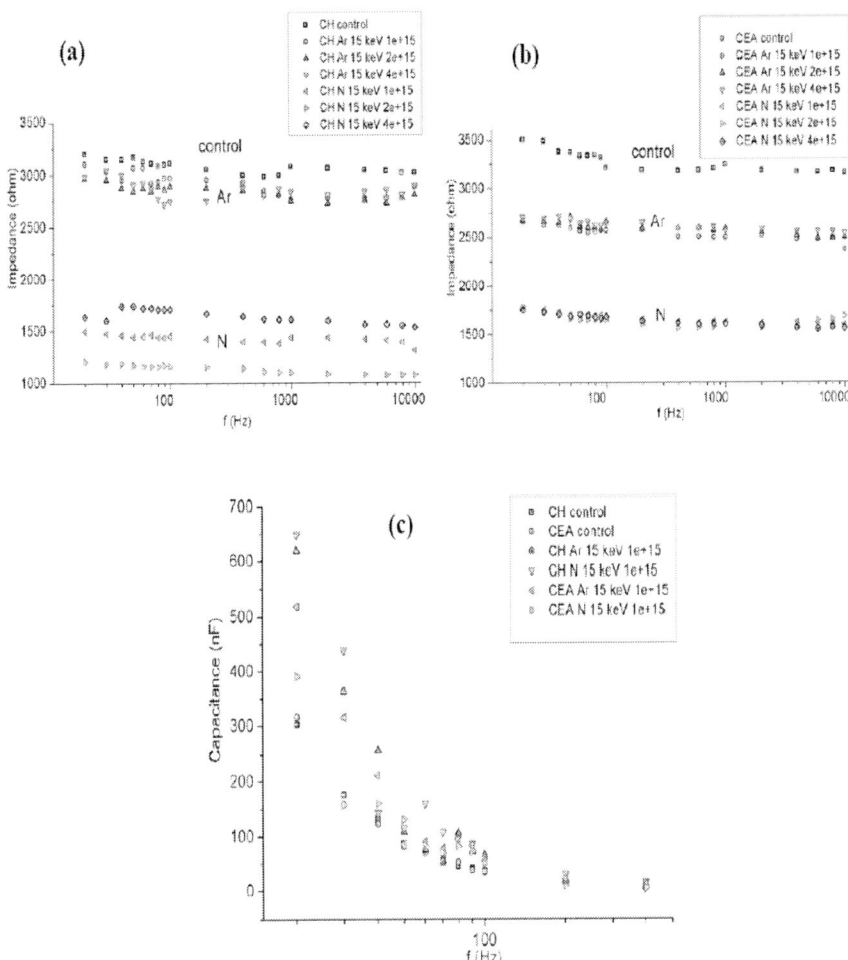

Figure 16. Membrane electrical properties as a function of testing frequency for different beam conditions (Reprinted from [31], Copyright (2009), with permission from Elsevier). (a) Impedance of chitosan membrane, (b) impedance of cellulose membrane, and (c) capacitance of each type of membrane. CH: chitosan. CEA: cellulose.

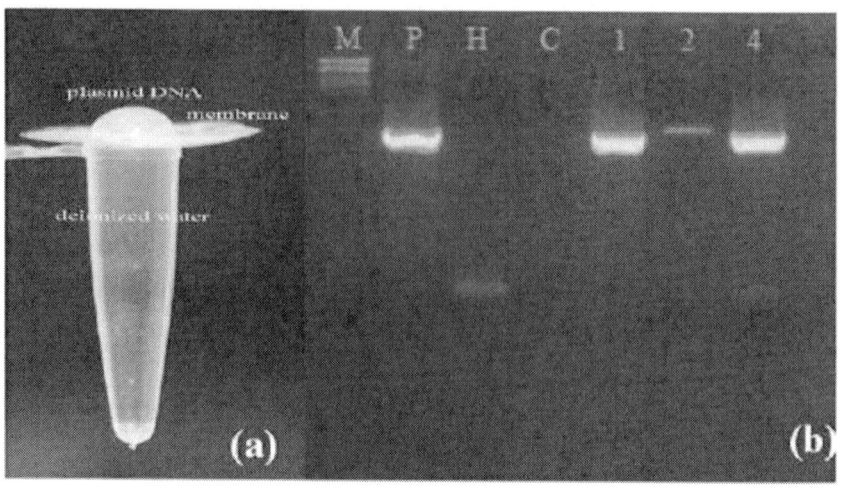

Figure 17. In vitro test of plasmid DNA transfer through the membrane (Reprinted from [31], Copyright (2009), with permission from Elsevier). (a) Photo of the experimental layout. The PCR tube is about 3 cm long. (b) Gel electrophoresis result. M: marker; P: plasmid DNA; H: deionized water; C: unbombarded membrane; 1, 2, 4: N-ion bombarded membranes with the beam conditions of 25 keV and 1, 2, $4 \times 10^{15}$ ions/cm$^2$, respectively.

Table 2. Plasmid DNA (pGFP) solution concentrations measured above the cellulose membrane and in the PCR tube under the membrane in the *in vitro* test of ion-beam induced DNA transfer through the membrane [31]. The ion bombardment conditions: nitrogen ions at 25 keV to the fluences as shown in the table. The percentage of transferring is calculated by a ratio of the DNA concentration measured in the PCR tube over the DNA concentration above the membrane

| Conditions | | DNA solution concentration (ng/µL) | % of transferring |
|---|---|---|---|
| DNA pGFP plasmid at top of the membrane (before transfer) | | 975.00 | - |
| DNA pGFP plasmid at bottom of the membrane (after transfer) | Unbombarded | 1.31 | 0.13 |
| | Bombarded with $1 \times 10^{15}$ ions/cm$^2$ | 10.08 | 1.03 |
| | Bombarded with $2 \times 10^{15}$ ions/cm$^2$ | 10.75 | 1.10 |
| | Bombarded with $4 \times 10^{15}$ ions/cm$^2$ | 7.92 | 0.81 |

## CONCLUSIONS

From our investigations on the ion beam induction of DNA transfer and related mechanisms, conclusions can be drawn as following.

- Ion beams with energy such that the ion range is approximately equal to the solid cell envelope thickness, at a certain range of fluence, are able to induce direct DNA transfer in the living cells.
- Due to special structures of the cell envelope, the ion energy and fluence required to induce DNA transfer can be sufficiently low.
- Ion beam bombardment of biological cells generally induces formation of micro/nano-crater-like structures in the cell envelope and they are speculated to act as pathways for exogenous macromolecules to transfer into the cell.
- Ion beam bombardment induced increase in surface roughness, decrease in the impedance and increase in the capacitance of the cell envelope favor DNA transfer, but change in hydrophilicity or hydrophobicity is not a key factor for the transfer.
- Compared with other physical methods for DNA transfer, such as electroporation, particle gun or microinjectile bombardment, microinjection and macroinjection, microlaser and ultrasonication, ion beam bombardment possesses certain advantages such as capability to treat a large number of various biological cells with high efficiencies, easy sample preparation for bombardment and biological treatment being a post process and thus convenient.
- Applications of low-energy ion beam bombardment induction of direct gene transfer have been practiced to improve local biological products.

*Chapter 4*

# ION BEAM INDUCED MUTATION

## 4.1. INTRODUCTION

Induction of mutation by external energetic particles stems from the idea that the particles penetrate through the coat of the germinating parts of biological organisms such as the embryo of a seed or the buds and then interact with the genetic substance in the cells of the germinating parts. The anatomies of the rice seed and the plant bud show that the embryos are normally deeply located inside the seed, e.g. the depth of the embryo of the rice seed from the seed coat surface is about several tens of micrometers, and the primordia of the bud is normally deeply buried beneath many bud leaves. In order to induce mutation, energetic ions need to interact with the embryo in the seed or the primordia in the bud through the materials that cover these germinating organs. In normal operation conditions, biological organisms are put in vacuum environment and subjected to ion beam bombardment. Instead of using high-energy (the ion energy higher than 100 MeV) ion beam irradiation, we applied low-energy ion beam of a few tens of keV to bombard the biological targets. Therefore, special measures should be taken. In the case of seeds, if the seed is big it is better to remove the seed coat above the embryo part beforehand to expose the embryo part directly to the ion beam for high-efficiency mutation induction. In case the seed is very small, if the embryo location is known it is better to orient the embryo part towards the ion beam; if the embryo location is unknown, the seeds have to be positioned in random for ion bombardment. In the case of tissues such as buds, the non-critical parts of the tissue should be well wrapped to prevent the water from evaporation to worsen the vacuum and to expose only the germinating part to the ion beam.

Here we introduce some examples of our successes in ion-beam-induced mutation.

## 4.2. MUTATION BREEDING OF RICE (*ORYZA SATIVA INDICA*)

Two varieties of local rice (*Oryza sativa indica*), purple glutinous rice (Kum Doi Sa Ket) [33] and Thai jasmine rice (KDML 105) [34], were chosen for the mutation induction purpose.

### 4.2.1. Purple Rice

Thai purple rice is native to Thailand. It has a typical rice shape, unique color and exotic taste and it is glutinous rice. Before bombardment, seed coats of the Thai purple rice were peeled out, and then the seeds were placed individually in a way that the embryos were faced to the ion incidence in the target holder which was made from copper for easy thermal transportation. The holder contained about 800 tiny holes whose sizes fit to the seed size (Figure 3b). Four such sample holders were installed to a holder stage which had four sides in the vacuum chamber of the high-beam-current ion implanter (Figure 3a). Nitrogen ion beam mixed with both atomic (N) and molecular ($N_2$) ions accelerated with high voltages of 60 – 120 kV, which gave ion energy of 30 – 60 keV for the majority of the nitrogen ions as soon as they were implanted, were used to bombard the seeds to ion fluences of $1 - 8 \times 10^{16}$ ions/cm$^2$. The operating pressure in the chamber was at an order of $10^{-4}$ Pa. After bombardment, the seeds were sown on moistened filter paper at room temperature under standard day light condition until germination. Then seedlings were transferred to grow in soil until the ripening stage. Both percentage of germination and survival were recorded. Phenotypic and genetic changes were observed in M1-M2 generations [33].

The results showed that the percentages of germination and survival of purple rice were decreased as increasing ion energies and fluences. This is certainly because of higher ion energy and fluence causing more damage in the cells. Rice seedlings with green pigment were observed in M1 generation while the wild type was purple (Figure 18). Seeds in the M1 generation were harvested and cultivated for M2 generations. Their phenotypes were divided into 3 groups: 1) the whole plant was still green, 2) only the stem was green while the leaf blade and sheath were purple, and 3) the whole plant was purple (Figure 19). It was also observed that the pigment of the seed coat and pericarp were not purple as was found in the wild type (Figure 20).

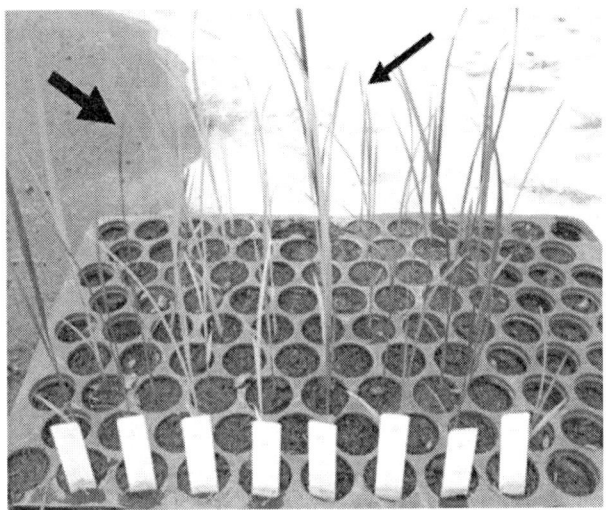

Figure 18. M1 generation of purple rice seedling bombarded by low energy ion beam. Thin and thick arrows show green seedling and purple seedling, respectively.

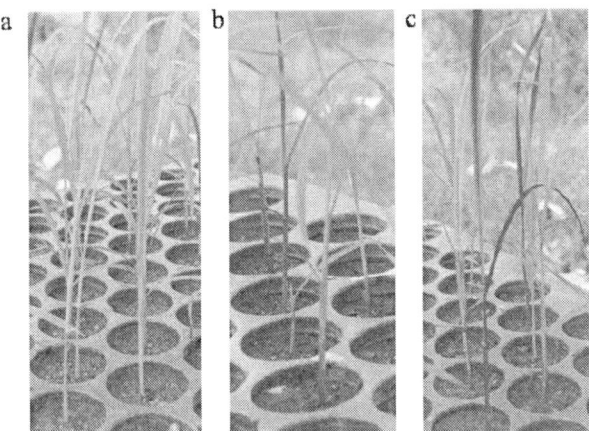

Figure 19. Pigment segregation in M2 generation of purple rice. Rice seedlings show green pigment (a), purple pigment at stem (b), and purple pigment (c).

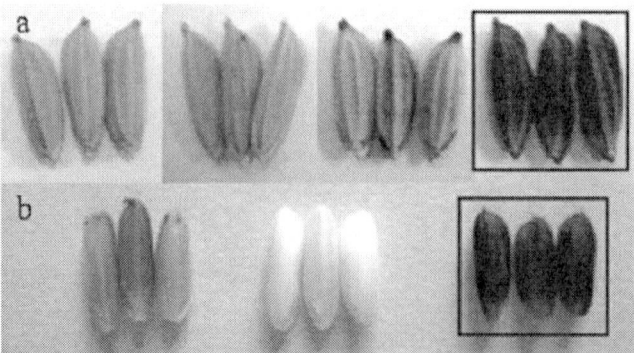

Figure 20. Phenotypes of M2 seeds of purple rice. Pigment of seed coat (a) and pericarp (b) were not dark purple as in wild type (shown in the square box).

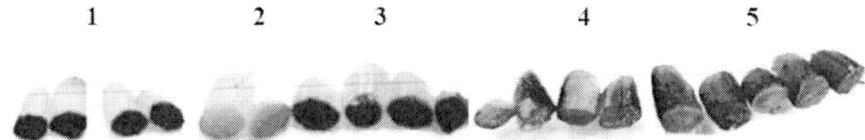

Figure 21. Starch content in purple rice seeds. After staining with aliquot of iodine solution the blue-dark color was observed in non glutinous rice seeds bombarded by nitrogen ions. Non glutinous Thai jasmine rice seed control (1), bombarded purple rice seeds with white pericarp from the same panicle (2 and 3), purple rice seed control (4) and non-glutinous black rice seeds (5).

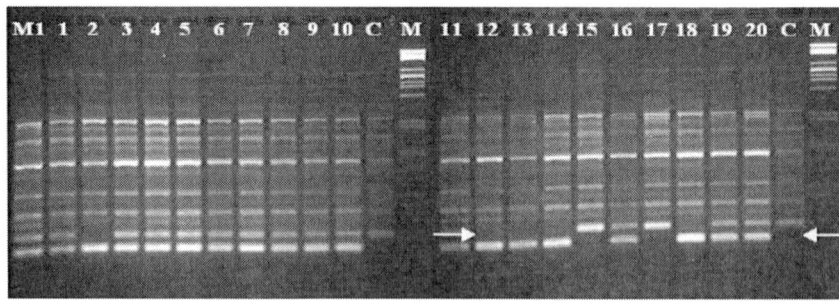

Figure 22. HAT-RAPD analysis in M2 generations of the ion-bombarded purple rice samples revealed by OPW09 primer. Lane 1-20: genomic DNA of M2 generation samples bombarded by N-ion beam. Arrows indicate the distinguished bands between the bombarded rice and the control. M1: genomic DNA of M1 generation. C: genomic DNA of the control. M: λ/ *Pst* I marker.

Rice starch consists of two components, amylose and amylopectin. Amylose is a planar of glucose linked mainly by α (1-4) bonds promoting the formation of helix structure, whereas amylopectin is highly branched polymer of glucose mostly found in plant. Amylopectin is responsible for the sticky quality of glutinous rice. Iodine was used to identify amylose property because its molecules fit inside the helix structure of amylose which then absorbs a certain known wavelength of light. In the presence of iodine to amylose a blue-black color could be seen, indicating the non-glutinous property of rice. After seeds were harvested from the M1 generation, in one of the samples bombarded with N-ion beam accelerated with a voltage of 80 kV to a fluence of $4 \times 10^{16}$ ions/cm$^2$, the starch content in the seeds showed a blue-black color versus the normal color (Figure 21). This indicated that low energy ion beam induced conversion from glutinous to non-glutinous rice. Using HAT RAPD (high annealing temperature – random amplified polymorphic DNA) marker revealed genotypic differences among rice samples in the M2 generation, indicating genetic modification occurring in their genome (Figure 22)

## 4.2.2. Thai Jasmine Rice

Thai jasmine rice (*Oryza sativa* L. cv. KDML 105) seeds were prepared for ion beam treatment. Fresh harvested seeds were heated at 49 °C in an oven system for 5 days in order to break rice dormancy. The rough seeds were then carefully husked to avoid any damage to the embryo tissues. The seeds were put in the sample holders, which had 800 tiny holes that fitted to the seed size and let the embryo part be exposed to the ion beam. Four sample holders holding the rice seeds were then installed onto a rotatable sample holder stage in the target chamber of the ion implanter. Thousands of the rice seeds were bombarded by nitrogen ions at various ion energy levels of 30 – 60 keV and ion fluences of $1 \times 10^{16} - 5 \times 10^{17}$ ions/cm$^2$. To prevent overheating of the rice seeds during ion bombardment, a pulse beam mode with a 10-to-10-second interval and water-cooling of the sample holder were adopted. After ion bombardment, the ion-bombarded seeds and controls were kept in moist conditions and then cultured in soil for 3 – 4 weeks. The seedlings were then transferred to grow as transplanted rice in soil in plastic pots as 1 seedling/pot. The cultivations were carried out in in-season cultivation (July – December) to screen for short-in-stature characters, and cultured in off-season cultivation (March – July) to screen for photoperiod-insensitive characters. Phenotypic and genomic variations in all growth stages, dates of flowering, dates of harvesting, plant height as culm length at harvesting day, etc., were recorded and analyzed [34].

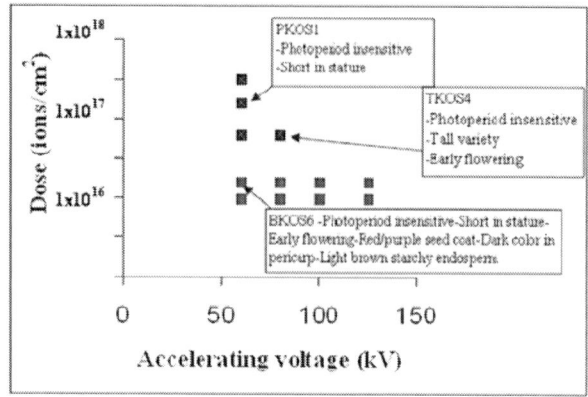

Figure 23. Induced mutations found in the rice grown from the KDML 105 rice seeds bombarded with nitrogen ions at various ion energy levels and ion fluences (doses) (Reprinted from [34], Copyright (2007), with permission from Elsevier).

Various mutants were obtained after screening as shown in Figure 23 [34]. Figure 24 shows the height statures of the rice before and after N-ion beam bombardment. The culm length of the shorter rice mutants is 67 cm and that of the taller mutants is 120 cm, whereas that of the control is 94 cm. Figure 25 shows various phenotypic variations [34].

Figure 24. Comparison of the height statures of the Thai jasmine rice KDML 105 mutants induced by N-ion bombardment with the control in M6 generation. The right one is the control, the two in the middle are the shorter mutants and the left one is the taller mutant.

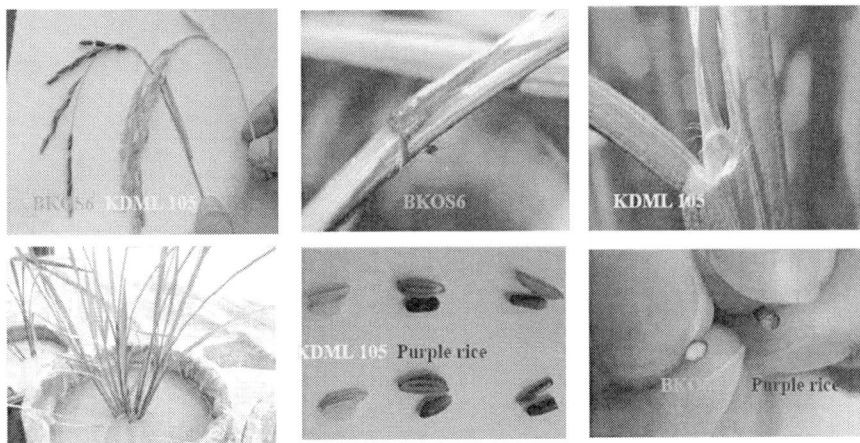

Figure 25. Various phenotypic variations found from M1-generation variation induced by N-ion bombardment. Here, KDML 105 is the control, BKOS6 is one of the variation samples induced by ion bombardment, and Purple rice is the local Thai purple rice as another control for comparison in the rice grain color.

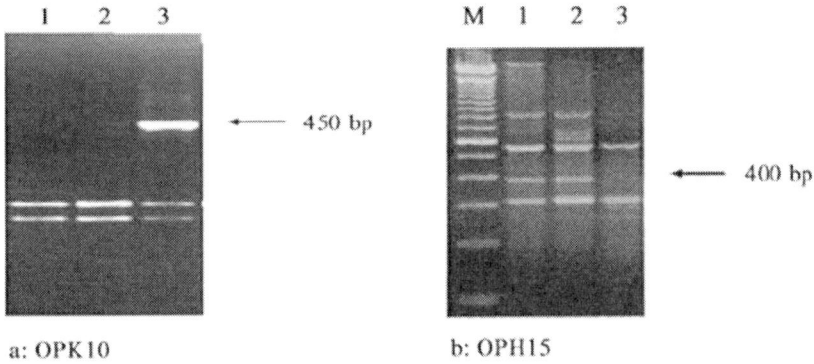

Figure 26. HAT-RAPD amplification products generated by OPK10 (a) and OPH15 primers (b) using KDML105 genomic DNA as templates from (1) fresh control, (2) vacuum control and (3) mutant BKOS6 (Reprinted from [34], Copyright (2007), with permission from Elsevier). M: the molecular weight marker lane. Arrows: the distinguished PCR products.

## Table 3. KDML 105 rice mutant lines obtained using ion beam biotechnology

| Rice label | Phenotypic variations |
| --- | --- |
| BKOS6 | Photoperiod insensitive, early flowering, short in stature, reddish dark brown to black in leaf sheath |
| TKOS4 | Photoperiod insensitive, early flowering, low tillering capacity, low % filled spikelets, low number panicles/plant, tall variety |
| PKOS1 | Photoperiod insensitive, big and long panicles, high % filled spikelets, high number of seeds in panicle, short in stature |

In DNA analysis, samples of leaf tissue were ground in liquid nitrogen to a fine powder and DNA extraction was performed following Doyle and Doyle method [35] but with minor modifications. Amplification of the DNA was carried out by the HAT-RAPD (high annealing temperature-random amplified polymorphic DNA) [36] technique. The amplification products were separated by electrophoresis and the result is shown in Figure 26. Additional DNA bands found in mutants were subcloned into plasmid pGEM-T easy vector (Promeca) and transformed into *E. coli* by electroporation and sequenced by the method of Sanger *et al.* [37]. The nucleotide sequences of the PCR products, designed BKPK10450 and BKPH15400, and their deduced amino acid sequences were compared with all sequences available in the GenBank database. The sequence analysis revealed [34]:

1. BKPK10450 belonged to a member of flavonoid 3'hydroxylase of *O. sativa japonica* with the highest identity of 60%. The flavanoid 3'hydroxylase is the enzyme that is involved in anthocyanin biosynthetic pathway. Anthocyanin pigments display color ranging from bright red/purple to blue. Color variation of the main red/purple in various tissues such as leaf sheath, collar, auricles, ligule, and pericarp, and light brown in starchy endosperm in BKOS6 might be induced from mutation in genes controlling the purple/red to blue.
2. BKPH15400 was 61 % identity to cytochrome P450 of *O. sativa japonica*. The BKPH15400 which is a member of cytochrome P450 may play significant role in biological systems, such as hormonal regulation, phytoalexin synthesis, as well as flower petal pigment biosynthesis, maybe resulting in the phenotypic variations.

Five KDML 105 rice mutant lines have been finally obtained and stabilized after M9 generation and three examples are shown in Table 3.

## 4.3. FLOWERS

### 4.3.1. Experiment

Seeds and buds of various local flowers such as petunia, rose, gerbera and chrysanthemum were bombarded by nitrogen ion beams at energies of 30 – 60 keV and fluences of $1 \times 10^{16} - 5 \times 10^{17}$ ions/cm$^2$ in vacuum [38]. In the case of bombarding buds, the stems that contained the buds were wrapped very well with plastic film, as shown in Figure 27, to prevent water in the stem from being continuously vaporized by pumping, which would break the vacuum condition, to only expose the buds to the ion beam. After ion beam treatment, the seeds or buds were cultured to grow.

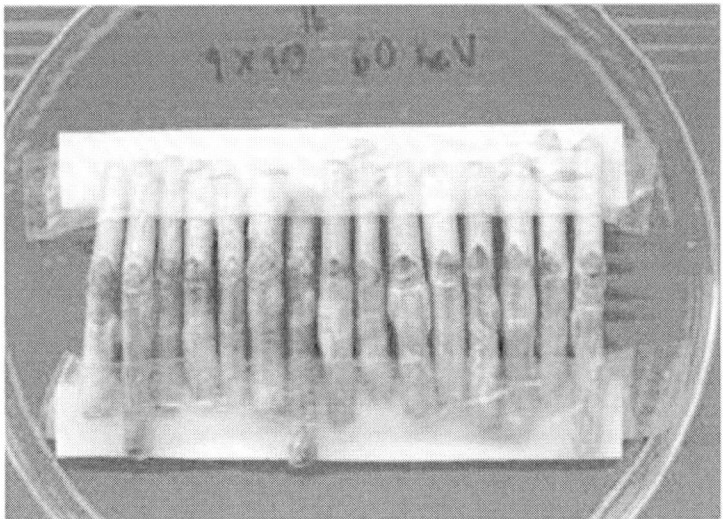

Figure 27. Photograph of prepared flower rose buds with the stems for ion bombardment.

Figure 28. Ion-beam-induced flower variations in the flower shape changes of petunia (a) control and (b) variation, rose Ingrid Bergman (c) control and (d) variation, gerbera spp. with (e) three headed and (f) fused flower, and impala lily flower control with five petals (right in (g)), a variation with four petals (left in (g)), and a variation with 6 petals (h).

## 4.3.2. Results

After screening, various variations were found with features of changes in the flower shape (Figure 28), color (Figures 29 and 30), petal shape and size (Figure 31), and petal color (Figure 32).

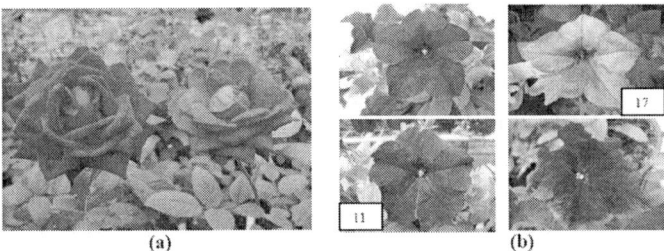

Figure 29. Ion-beam-induced flower variations in the flower color changes. (a) Rose Ingrid Bergman control (left) and variation (right). (b) Petunia control (up-left) and variations (others).

Figure 30. Variations in chrysanthemum flower color intensity. (a) Paler than (b) control, (c) darker than control, (d) variegated color, and (e) varied color from pink to bronze.

Figure 31. Variations in the petunia flower petal shape and size. Up-left: control, and others: variations.

Figure 32. Ion-beam induced variation in Gerbera flower petal color. The natural control has monocolor of red.

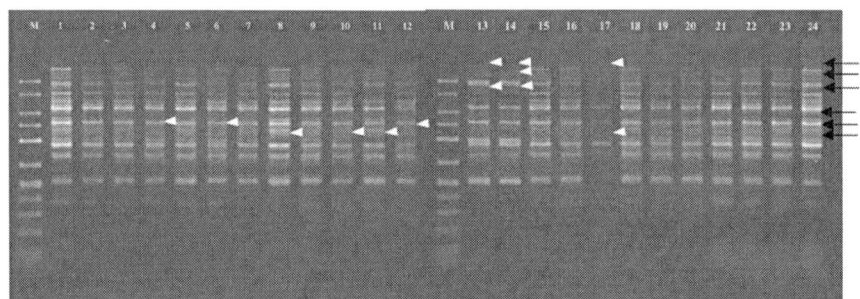

Figure 33. DNA banding patterns of the control and the variants of petunia using OPA 04 as a primer. The black arrows point the polymorphic bands and the white arrows point the bands from the variants, which are different from those of the control. Lane 1: control. Lanes 2 – 24: variants, among which some are denoted by the numbers in Figures 29 and 31.

DNA analysis by RAPD was carried out for all of the mutants. For example, twenty-four variants of petunia were selected for identification at the molecular level and the results are shown in Figure 33. In the analysis, eleven primers, OPA 01 OPA 04, OPA 05, OPA 07, OPA 08, OPA 09, OPA 10, OPA

11, OPA 14, OPA 15 and OPA 16 (Operon Technology Alamada, USA), were used to characterize polymorphisms among the 24 petunia variants (some of which are shown in the figures above). Four of the eleven primers demonstrate DNA polymorphic bands in the range of 550-3120 base pairs. Primer OPA 04 shows 13 bands, 7 of which are monomorphic bands and the others polymorphic bands. Banding patterns of 8 variants (lane 4, 6, 8, 10, 11, 12, 13, 14 and 17) are different from that of the control.

## 4.4. VEGETABLES

Seeds of two vegetable species, cucumber (*Curcumis sativus*) and mung bean (*Vigna radiata*) were bombarded by low energy ion beam.

### 4.4.1. Cucumber

The variety of cucumber (*Curcumis sativus*) used in this experiment was Gy variety whose reproductive organ was gynoecious only. Dry seeds were bombarded with 80-kV accelerated mixed nitrogen ions to fluences of 2, 4, 8 × $10^{17}$ ions/cm$^2$. After bombardment, the seeds were grown in conventional soil conditions. Effects of ion beam on the seed germination and growth were assessed in terms of both quantitative and qualitative traits. The percentages of the germination of the bombarded seeds with the fluences mentioned above were 92, 91 and 88, respectively, while 92 for the control. The percentages of the survival of the bombarded seeds were 99, 99 and 89, respectively, and 99 for the control. The results indicated that ion bombardment had slight effect on its germination and survival. However, the control group possessed more height and growth rate than the ion-bombarded group of cucumber significantly. In addition, about 3% of the plants grown from the seeds bombarded with the fluence of 2 × $10^{17}$ ions/cm$^2$ showed their size of the female reproductive organ (ovary) greater than that of the control and some were transformed into the male reproductive organ (Figure 34a). Crossing pollination between the greater organ to the male organ of the wild type resulted in a short fruit size but more fresh meat and less seeds (Figure 34b).

Figure 34. Effect of ion beam on the reproductive organ of cucumber. (a) The size of the female reproductive organ (ovary) induced by ion bombardment (right) was greater than the control (left). Arrows indicate the ovaries. (b) Cucumber flower without ovary induced by low energy ion beam (male cucumber). (c) Size comparison of the whole fruit of cucumber between control (left) and hybrid (right). (d) Fresh meat of control (left) and hybrid (right).

### 4.4.2. Mung Bean (*Vigna radiata*)

Seeds of local Mung bean with the embryo parts facing to the ion beam were irradiated by nitrogen ion beam accelerated with 100, 80, 60 and 30 kV at the same fluence of $1 \times 10^{16}$ ion/cm$^2$. After ion bombardment, the seeds were grown in normal soil conditions. Effects of ion beam on the seed germination and growth were assessed in terms of both quantitative and qualitative traits. The percentages of the germination of the seeds bombarded with the ion acceleration voltages mentioned above were 58.00, 64.41, 62.00 and 63.00, respectively, and 88.06 for the control. The percentages of survival were 89.66, 84.21, 91.94, 96.83, respectively, and 98.31 for the control. The results showed that the control group possessed more height, leave number, growth rate and dry mass than the ion-bombarded group of Mung bean significantly. From the observations, it seemed that the nitrogen ion beam at energy of 30 keV with the fluence of $1 \times 10^{16}$ ion/cm$^2$ was the most effective to variation of the bean. Qualitatively, variations of abnormal stems (swollen shoot) and leaves (wrinkled, small) were identified as shown in Figure 35.

Figure 35. Ion-beam-bombardment induced variations in green bean. Compared with the control (a), it is seen that in (b), the leaves are rolled up and shrunk (observed in 30% of the total seedlings), and the stem bent; in (c) there grow double stems, observed at only one seedling among 100 (it should be only a single stem); in (d) there grow a branch (it should not have a branch) as well as rolled and shrunk leaves; in (e) the stem becomes fairly thick, observed at only one seedling among 100 (it should be thin as shown by the control).

## 4.5. GENE CLONING

Ion beam induction of mutation can be applied for gene cloning. Biological organisms have so many functions, which are controlled by genes. Some have been known, but some unknown. Because ion beam can induce broad-spectrum mutations, it is convenient to apply ion beam to induce variations, among which there may be variants to be investigated for identification. Here is an evidence of gene cloning by application of low-energy ion beam [39].

Anthracnose, caused by the fungus *Colletotrichum* sp., is one of the important diseases affecting flowers. The use of natural antagonists has recently been applied for biological control. In the study, the target material was *Bacillus* (*B.*) *licheniformis* (a kind of bacteria) isolated from hot springs in Chiang Mai, showing its activity to suppress conidia germination of the fungus and reduce symptom caused by the disease in flower plants. N-ions at energy of 28 keV were chosen to bombard the bacteria to a fluence range of $1 - 10 \times 10^{15}$ ions/cm$^2$. After ion bombardment, bacterial variants were screened and one of the variants with losing its antagonistic property was obtained (Figure 36). For this variant, DNA fingerprint analysis was carried out. Polymorphic bands were detected between the bombarded bacteria and the wild type by HAT-RAPD (high annealing temperature – random amplified polymorphic DNA) technique [40]. The results revealed that the fingerprint profiles of the bombarded bacteria and the wild type were different (Figure 37). The additional band presented in the control was subcloned into pGEM-T easy vector and sequenced. Partial sequence analysis revealed that this fragment was a gene encoding enzyme lipase. Regarding to the lipase gene sequence, a pair of specific primer was designed from the *B. licheniformis* database to amplify the entire sequence of the gene using its genomic as the template.

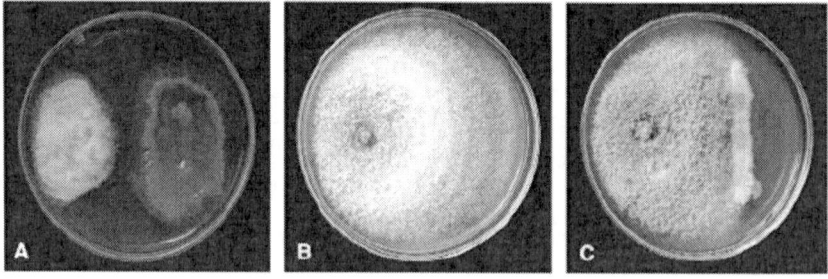

Figure 36. Antagonistic effect in vitro by dual-culture method: (A) *B. licheniformis*; (B) control; (C) mutation (MBL1).

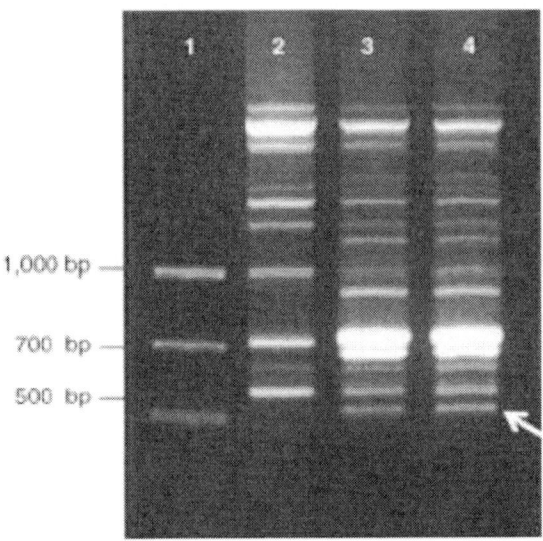

Figure 37. Amplification products of the bombarded bacterium using primer OPAH 10. (1) marker; (2) bacteria (MBL1) losing antagonistic activity; (3) bacteria having the same ability for inhibition; (4) control (the arrow indicates additional band subcloned into pGEM-T easy vector).

In order to determine the expression of the lipase gene in yeast, the entire lipase gene was subcloned into pYES2 and named pYES2-LicLip, then transferred into yeast cell via low-energy ion beam. The total protein in the volume 10 µl from *S. cerevisiae* (non transferred), *S. cerevisiae* transferred with pYES2, *S. cerevisiae* transferred with pYES2 and induced, *S. cerevisiae* transferred with pYES2-Liclip, *S. cerevisiae* transferred with pYES2-Liclip and induced, *B. licheniformis* and *B. licheniformis* induced colonies were spotted onto 0.5-cm diameter sterile paper (the yeast cells were induced by supplementing with galactose fertilizer while *B. licheniformis* was induced by co-cultivating with the disease fungi). The paper was placed in an assayed PDA plates, containing *Colletotrichum musae*, *Pyricularia grisea* or *Fusarium oxyaporum* which were plant disease fungi. Subsequently, the assayed PDA plates were incubated at room temperature until clear zones were observed (about 3 days for each of the three replicates). The result showed that only the transformed with pYES2-Liclip induced colonies and *B. licheniformis* induced colonies were presented the clear zone (Figure 38). The size of clear zone was showed in Table 4. The wide clear zone of antifungal activity from

*B. licheniformis* was presented by comparing with the diameter of protein from *S. cerevisiae* transformed with pYES2-Liclip [39].

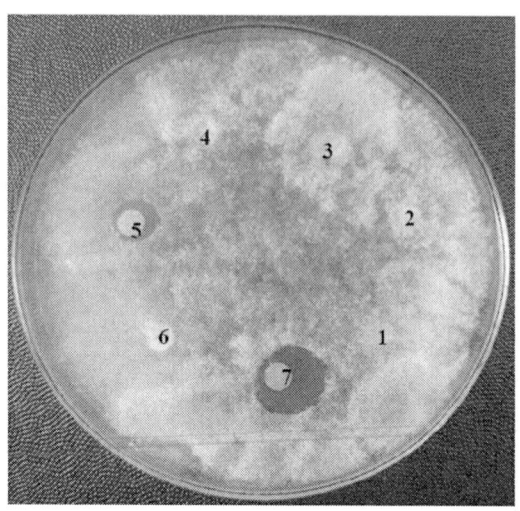

Figure 38. The analysis of antifungal activity from protein extracts measurements, with plant disease fungi (*Colletotrichum musae*) (Reprinted from [39], Copyright (2009), with permission from Elsevier). 1: *S. cerevisiae*, 2: *S. cerevisiae* transformed with pYES2, 3: *S. cerevisiae* transformed with pYES2 and induced, 4: *S. cerevisiae* transformed with pYES2-Liclip, 5: *S. cerevisiae* transformed with pYES2-Liclip and induced, 6: *Bacillus licheniformis*, and 7: *Bacillus licheniformis* induced.

**Table 4. The analysis of antifungal activity from protein extracts**

| Microbes | Diameter of the clear zone (cm) | | |
|---|---|---|---|
| | *Colletotrichum musae* | *Pyricularia grisea* | *Fusarium oxyaporum* |
| *S. cerevisiae* | - | - | - |
| *S. cerevisiae* transformed with pYES2 | - | - | - |
| *S. cerevisiae* transformed with pYES2 and induced | - | - | - |
| *S. cerevisiae* transformed with pYES2-Liclip | - | - | - |
| *S. cerevisiae* transformed with pYES2-Liclip and induced | 0.55 | 0.72 | 0.50 |
| *Bacillus licheniformis* | - | - | - |
| *Bacillus licheniformis* induced | 1.67 | 2.23 | 1.37 |

## 4.6. DISCUSSION ON MECHANISMS

How low-energy ion beam can induce mutation is a puzzle. For example, PROFILE-Code calculated N-ion depth distribution in the plant cell wall in vacuum shows that the ion range is about 220 nm under the ion implantation conditions of energy of 50 keV and a fluence of $5 \times 10^{16}/cm^2$, as shown in Figure 39. In the figure, the cursor at the depth of 400 nm points out the ion density to be about $6.4 \times 10^7/cm^3$ as shown at the up-left corner box, or 640 ions/$cm^2$ in a region of 100 nm around this depth, or only about $10^{-14}$ the fluence applied. However, the experimental fact was that the variation rate was about 1/1000 and the material thickness above the embryo was much greater than 400 nm. The question is then how less than $10^{-14}$ applied ions can induce variation in a rate of $10^{-3}$.

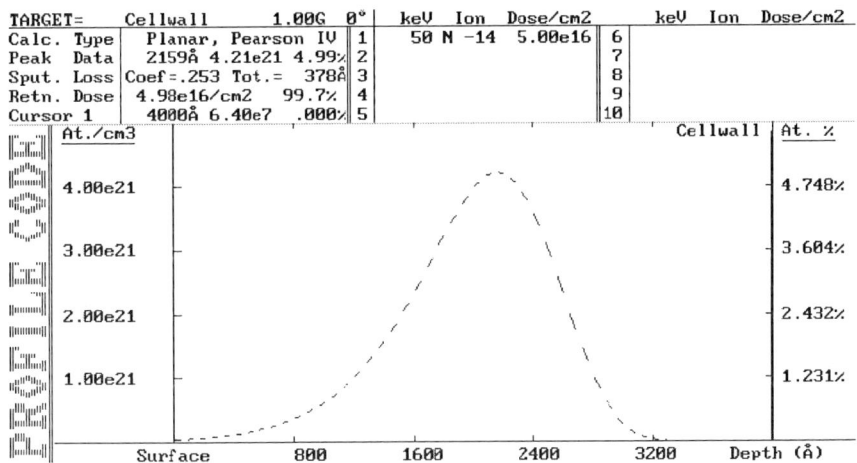

Figure 39. PROFILE code calculated N-ion distribution profile in the plant cell wall. As shown in the up-left corner, the cursor is put at 4000 angstrom, where the ion density is only $6.40 \times 10^7$ ions/$cm^3$.

It is noted that the computer calculation assumes a homogenously-structured cell wall material to be the target as it lacks of sufficient information on the real compositions and structures about the biological materials including the seed coat and the embryo coat and membrane above the embryo. The reality may be far deviated from the assumption. For example, the plant seed coat can be considerably porous [41] and thus the stopping power of the seed coat to the incident ions can be significantly lower than that of a homogenously dense seed coat material as assumed by the computer program. Furthermore, we have found that there exist a number of cracks in the plant seed coat or the embryo membrane, as shown in Figure 40. These cracks may be channels for incident ions to penetrate deeply without stopping into the embryo.

The discussion above only refers to possible direct interaction from ion bombardment with the embryo. Ion bombardment also produces indirect effects which may play roles in inducing mutation. The indirect effects include heat generated by ion beam bombardment caused energy deposition, secondary electron emission due to collision between the incident ions and the target atoms, X-ray emission caused by the incident-ion-excited electrons and free radicals directly produced by the incident ions impacting biological molecules. All of these secondary products may interact with the genetic substance to cause mutation. Some of these have primarily been theoretically discussed [42], but further experimental investigations are needed to provide evidence to confirm or negate the secondary effects.

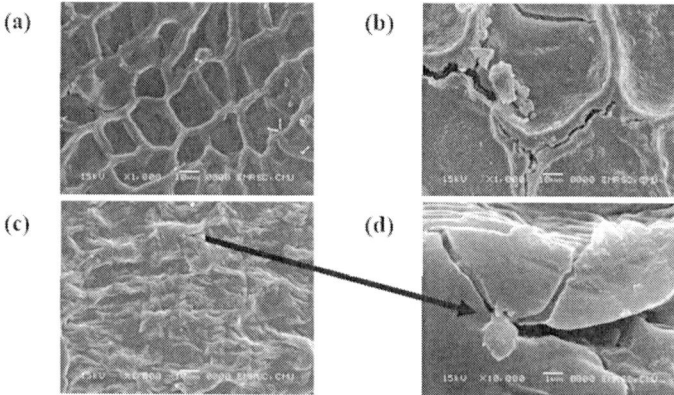

Figure 40. SEM microphotographs of the surfaces of (a) rice seed embryo, (b) flower petunia seed coat, (c) green bean seed embryo, and (d) zoom of the crack from (c).

## 4.7. VERY LOW-ENERGY AND LOW-FLUENCE ION BOMBARDMENT EFFECT ON NAKED DNA

The range of such low-energy ions in plant cell materials has been predicted to be several-fold that calculated for compressed solid of the cell materials [23]. Furthermore, as discussed above, the cell envelope structure is highly porous and cracked. Thus, it is statistically possible for a very small portion of ions able to penetrate through the cell materials including the cell envelope and cell inner substance that cover the cell nucleus to reach DNA. In this case, both energy and fluence of these ions must be very low. A question raised is whether the few very low energy ions can still induce structural change in DNA. A good way to study this issue is to use low-energy and low-fluence ions to directly bombard naked DNA in vitro to simulate the final-step interaction between the ions and DNA. There have been some investigations done on various low-energy radiations including low-energy ion bombardment of DNA [43-48]. The results showed that low-energy ions could produce plasmid DNA strand breaks. However, our investigation was the first attempt to use very low energy nitrogen ions, which were the ion species most popularly applied for ion beam mutation practice and also the abundant element in DNA, and compare effects from N ions with inert and heavy Ar ions [49].

### 4.7.1. Theoretical Estimation of the Low-Energy Ion Range in Naked DNA

In order to know in what range the low-energy ion can interact with naked DNA, we have developed an approach for the calculation of ion range, using a simplified mean-pseudo-atom model of the DNA target [50]. According to the model, a DNA molecule has a mean chemical composition of $H_{12}C_{10}N_4O_7P$ with a mean atomic number of 5.03 and a mean atomic weight of 9.74. For different forms of DNA, e.g. A-DNA, B-DNA and Z-DNA, the mass density ranges from about 0.6 to 1.1 g/cm$^3$. Based on the ion stopping theory [51], for the case of low-energy ($E \leq$ a few keV) ion implantation into DNA, the nuclear stopping $S_n$ falls in the low reduced energy regime, which gives a cube-root energy dependence of the stopping ($E^{1/3}$)

$$S_n = AE^{1/3}, \qquad (1)$$

where

$$A = 4\pi a_{TF}^{4/3} (Z_1 Z_2 e^2)^{2/3} \left(\frac{M_1}{M_2}\right)^{1/3} \left[\frac{M_1 M_2}{(M_1 + M_2)^2}\right]^{2/3}. \qquad (2)$$

Here, $a_{TF}$ is the Thomas-Fermi (TF) screening radius, $Z_1$ and $Z_2$ are the atomic numbers of the incident ion and the target material element, respectively, $M_1$ and $M_2$ are the atomic weights of the incident ion and the target material, respectively, and $e$ is the electronic charge. Calculation formulae of the ion range in DNA are then obtained to unify the relevant calculations. Our calculation indicates that the low-energy ion range in DNA is in the nanoscale and sensitive to nanometer. For example, the projected range of 4-keV Ar ion in the nature form B-DNA is 7.5 nm, 1-keV C ion about 5 nm and 60-eV H ion about 1.5 nm. Upper limits of the ion energy as a function of the atomic number of the bombarding ion species are proposed for the low-energy case to hold. Comparison of the results of this approach with the results of some widely used computer simulation codes such as SRIM [52], PROFILE [23] and PRAL [53] and with results reported by other groups indicates that our approach provides convincing and dependable results.

### 4.7.2. Experiments

The experiments had two parts, one using ions from the beam-line accelerator (the bioengineering beam line as mentioned before) and the other using ions from plasma immersion ion implantation (PIII).

#### *4.7.2.1. Sample Preparation*
An initial sample of plasmid pGFP (plasmid green fluorescent protein, 3344 base pairs) was purchased from Clontech. The sample was replicated following transformation into *Escherichia coli* (*E. coli*) and subsequently extracted and purified using a QIAGEN® Plasmid Purification kit according to the manufacturer's protocol. The plasmid DNA was dissolved in sterile, de-ionized water resulting in a plasmid concentration of 1 µg/µl. The solution was divided into aliquots and later diluted in water as necessary.

#### *4.7.2.2. Ion Bombardment*

**(I) Ion Beam**
Aliquots of 1-µl plasmid DNA solution (containing 100 µg DNA unless otherwise indicated) was deposited on sample pots of a sample holder. The holder, as shown in Figure 41(a), was designed and made in special considerations. It was made from glass for bio-cleanness, and a set of glass tubes, instead of a dish which was found easily cracked by beam heating. On

the tubes, pots, each with 5 mm in diameter and 5 mm in depth, were glass-worked in separation to prevent DNA molecules from jumping out due to beam sputtering effect. The DNA samples in the pots were then dried by heating to 60 °C for 5 minute before bombarding. Nitrogen ions at 2.5 keV and argon ions at 5 keV (both having almost the same ranges in most materials) to fluencies of 3, 6, 9 × $10^{13}$ ions/$cm^2$ bombarded the naked plasmid DNA with the bioengineering-specialized ion beam line. In each ion beam bombardment, one condition was normally applied to three samples and a group of three samples which were covered with a stainless steel plate in the neighbor of the ion bombarded samples were used as the vacuum or internal control.

## (II) PIII

Aliquots of 9 µl plasmid DNA solution were deposited in holes of a stainless steel sample holder. The holder was about 5 cm in diameter and had nine holes on it with each in a size of 5 mm in diameter and 5 mm in depth, as shown in Figure 41(b). A hole containing the vacuum control was covered by carbon tape. The samples were first dried in laminar flow and then placed in the PIII chamber which was then pumped to a pressure $< 5 \times 10^{-5}$ torr. We used argon or nitrogen plasma for immersion ion implantation with bias voltages of 0 and 2.5 kV and fluences of 0, $10^{11}$, $10^{12}$ and $10^{13}$ ion/$cm^2$, respectively. When no bias was applied, the sample holder was either grounded or not grounded. The plasma was generated with 50-watt radiofrequency (RF) power and operated with a frequency of 50 Hz and a pulse length of 10 µs.

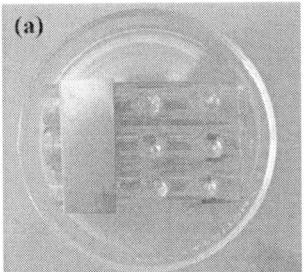
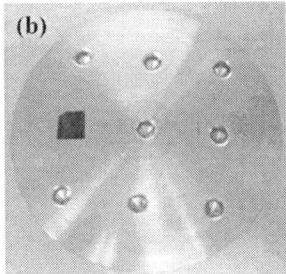

Figure 41. Naked DNA sample holders for (a) ion beam bombardment (Reprinted from [49], Copyright (2009), with permission from Elsevier), and (b) PIII.

### 4.7.2.3. Post-Bombardment Biological Treatment and Analysis

After bombardment, the samples of both control and bombardment were then individually recovered in 10 µl of de-ionized water for dilution and divided two parts for analysis. One part was stored at −20 °C for transfer into *E. coli* and others were then added to with 2 µl of gel loading buffer (0.25% (w/v) bromophenol blue, 40% (w/v) sucrose). The samples added with gel and also the natural or solution control and digested control samples were then loaded onto a gel electrophoresis apparatus. This gel was run at a constant voltage (100V cm$^{-1}$) for about one hour. Images of the gels were captured using UV-transilluminator for viewing DNA in agarose gels and digital camera for capturing. Fluorescence intensity plots were obtained using the Scion Image for Windows. There were totally four types of samples analyzed in electrophoresis: (1) ion bombarded, (2) vacuum or internal control, (3) natural or solution control, and (4) digested control. The internal control samples were subjected to the same preparation, vacuum and collection procedures as the bombarded samples but not subjected to bombardment. The solution control consisted of the same quantity of plasmid as the internal controls but it was not subjected to the deposition, vacuum and collection procedures. The digested control consisted of a sample of pGFP digested with restriction enzyme *EcoRI* (Sigma) in order to act as a marker for full length linear plasmid. The intensity corresponding to each form in the electrophoresis was quantified by integrating the area under the corresponding peak [44,54,55].

The ion-bombarded DNA sample which was stored at −20 °C was transferred into *E. coli* JM109 competent cells. And the transformants were screened on culture media containing 200 µg/ml IPTG (isopropyl *β*-D-thiogalactoside) for inducing gene expression. After overnight incubation at 37 °C, white colonies were picked out and then subcultured for 2 – 3 times in new plates for their purity.

## 4.7.3. Results and Discussion

### 4.7.3.1. Ion Beam Bombardment

The results from the electrophoresis analysis are shown in Figure 42 [49]. It is known that when a single-strand break (SSB) is induced, DNA converts into a relaxed form, and when a double strand break (DSB) or multiple DSBs are produced, DNA converts to a linear full-length form or fragments. From the figures, it is clearly seen that upon the very low-energy low-fluence ion bombardment both relaxed and linear forms are produced and hence SSB and DSB indeed occur. It is noticed that in the vacuum controls, the relaxed form is dominantly produced, indicating vacuum effect on DNA SSB. The changes in the amounts of the DNA forms as increasing the ion fluence are found related to ion species. As increasing the ion fluence, the amount of the original DNA supercoiled form decreases for the N-ion bombardment case but does not much change for the Ar-ion bombardment case; the amount of the relaxed form is almost stable for the N-ion case but slightly increasing for the Ar-ion bombardment; the amount of the linear form increases for the N-ion case more than for the Ar-ion case (Figure 43). This comparison indicates that nitrogen ions, even with lower energy than that of argon ions, are more effective in producing double strand breaks and thus more capable to induce *GFP* gene mutation than argon ions. This result seems to be conflict with common knowledge that predicts higher-energy and heavier ions able to produce more damage than lower-energy and lighter ions. Whether more physics and biology are involved is being further investigated. One hypothesis is that because DNA contains much nitrogen at the nitrogenous bases, externally introduced nitrogen will have intimate interaction with the original nitrogen so that more effects can be produced.

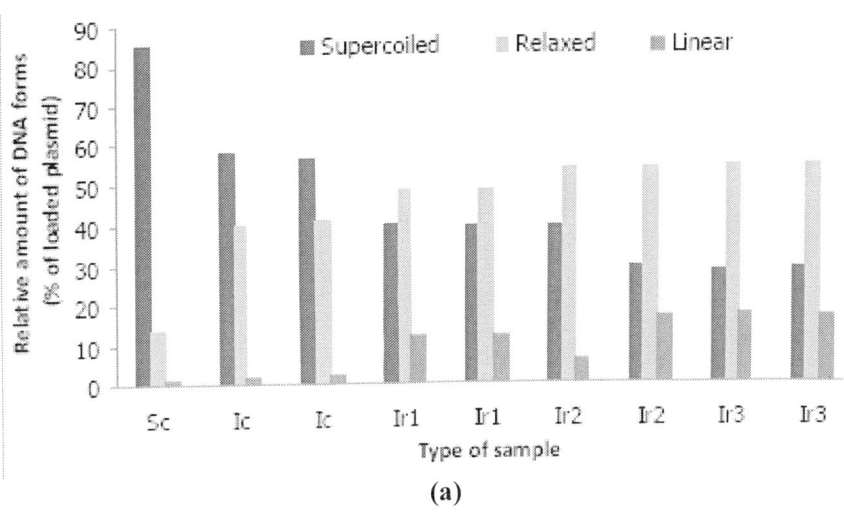

(a)

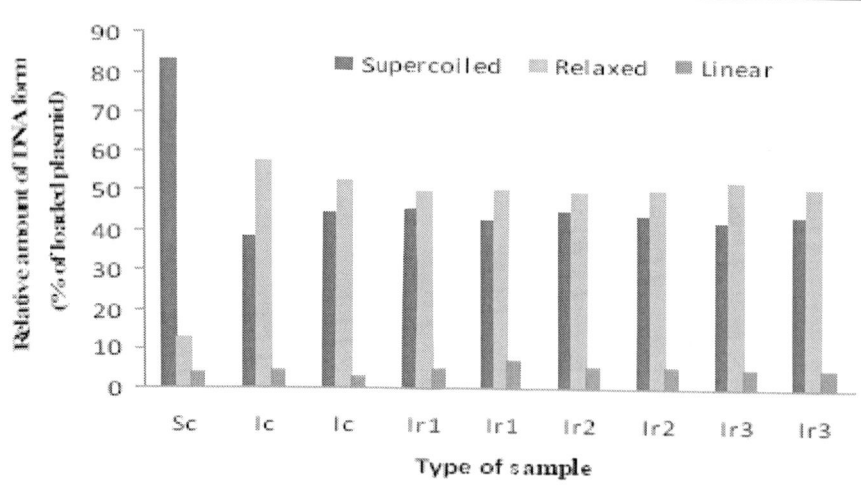

(b)

Figure 42. Quantified electrophoresis results (Reprinted from [49], Copyright (2009), with permission from Elsevier). The samples were separated on 1.4% agarose gel and visualized by ethidium bromide staining. Keys: Sc-solution control, Dc-EcoRI digested control, Ic-internal control, Ir-irradiated with fluences of $3 \times 10^{13}$ (Ir1), $6 \times 10^{13}$ (Ir2), and $9 \times 10^{13}$ (Ir3) ions/cm$^2$. (a) N-ion bombardment. (b) Ar-ion bombardment.

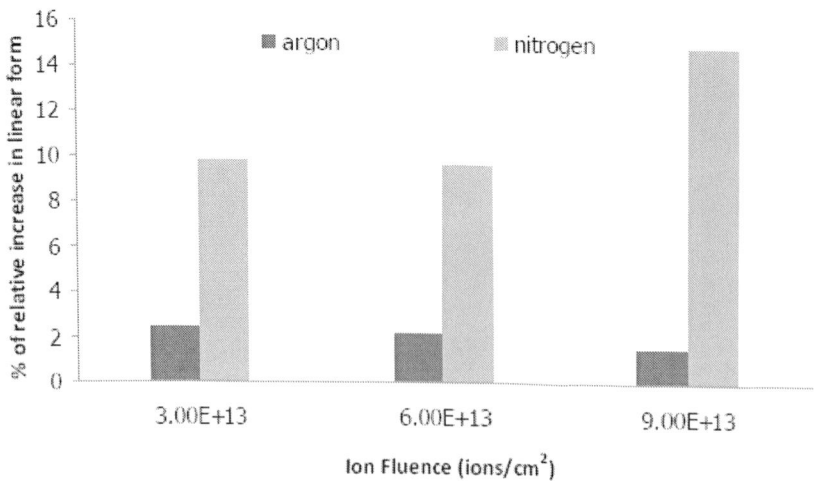

Figure 43. A comparison of the linear-form plasmid DNA between Ar-ion and N-ion bombardments (Reprinted from [49], Copyright (2009), with permission from Elsevier). The data are means of many sets of the electrophoresis results.

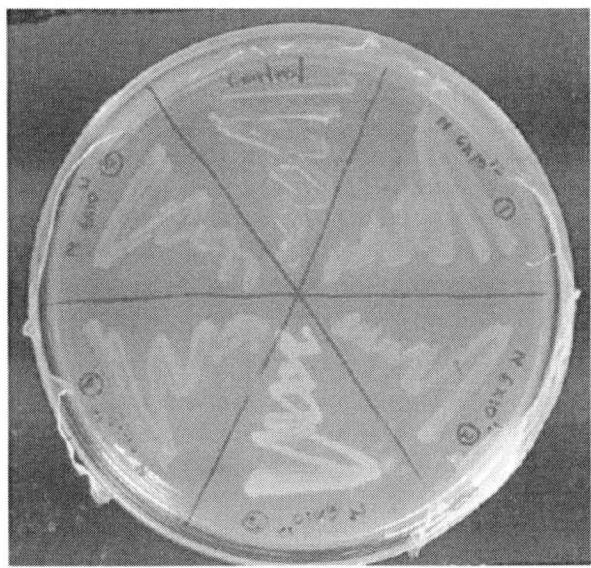

Figure 44. UV observation of plasmid containing green fluorescence protein (pGFP) transferred into *E. coli*. White (original bacterial color) colony indicates pGFP damaged (nonfunctioning) and thus mutated, while green colony is non-mutant.

The result of DNA transfer in *E. coli* showed that green (non-mutant) and white (mutant) colonies were produced. The white colonies were picked out and plated on plates again to check for their purity as shown in Figure 44. The appearance of white colonies that are the evidence of the *GFP* gene damaged and thus not functioning confirms that low-energy ion beam bombardment indeed induced DNA modification. Our gene sequencing showed that the sequences of the *GFP* gene in the mutants induced by both Ar-ion and N-ion bombardments were similar to that of the *GFP* in the control. This means that the *GFP* gene is not mutated. Therefore, the mutation may only be attributed to the Lac promoter, because *GFP* is expressed under the Lac promoter control to produce a chain of several additional amino acids, including the first five amino acids of the lacZ protein.

Note that the ion fluence was in the order of $10^{13}$ ions/cm$^2$, about three orders lower than that applied for normal ion bombardment induction of mutation. This is corresponding to, on average, only one ion reaching DNA per 1000 incident ions. This low fluence means about 1/10 ions impinging on DNA per n m$^2$. It is well known that the DNA chain is about 2 - 3 nm wide and one nucleotide unit is 0.33 nm long [56], or about 1-nm$^2$ area for a nucleotide unit. So, on average, every 10 units are bombarded by only one ion. This low probability of impingement has already caused significant changes in the DNA structure, demonstrating how sensitive and effective the low-energy ions are capable of inducing mutation.

### 4.7.3.2. Plasma Immersion Ion Implantation (PIII)

Vacuum effect on damage in DNA and subsequently induced mutation of DNA-transferred bacteria *E. coli* was first checked. As the plasmid DNA pGFP produced the green fluorescent protein which would be seen in green color under ultra violet (UV) light, green color observed in pGFP-transferred living materials was the indicator of the presence of the functioning pGFP DNA, whereas no green color shown should be the indicator of the DNA modified or damaged. No mutation was found from the *E. coli* transferred with plasmid DNA pGFP which was exposed to vacuum at a pressure of $10^{-3}$ Pa up to one hour. In fact, under all of the conditions applied (varied low pressures and exposure time lengths), the DNA-transferred *E. coli* all showed green. This result demonstrates that certain long-time exposure of DNA to vacuum basically has no effect on mutation.

In order to check effect from only plasma (without using bias) on bacterial mutation, we placed the DNA samples in the vacuum chamber with argon or nitrogen plasma generated from RF power input but without bias. Two conditions were applied, namely grounding the sample holder and not grounding. In the former case, the ions with only the thermal energy bombarded the DNA, while in the latter case, the ions only "blew" the DNA with the thermal energy. From both conditions, no mutation in the bacteria was found. This result means that only with the ion energy of an order of eV, mutation cannot be induced within the treatment time periods.

With applying the bias, mutation of *E. coli* transferred with the ion-bombarded DNA, indicated by the white bacterial colonies, was observed in all PIII conditions. Purification of the white colonies with picking up the colonies to grow in culture media LB exhibited all cells in green, demonstrating the white colonies not contaminated but really mutated. The maximum energy of Ar ions was 2.5 keV and the energy of most N ions was about 1.25 keV (as most N ions were molecular) in our conditions. Therefore, these very low energy ions were able to induce bacterial mutation directly.

The DNA forms under various conditions were analyzed with gel electrophoresis (Fig. 5) and quantified based on the fluorescence intensity of the band as shown in Figure 45. The results clearly show that the supercoiled form decreases while the relaxed form increases as the DNA is treated more and more. Vacuum can cause certain SSBs and the relaxed form increase by about 50% compared with that of the natural control. But, the ion bombardment considerably induces more SSBs with about 200% increase in the relaxed form for the Ar-ion case and 125% increase for the N-ion case at the lowest fluence and the relaxed form further increases as the ion fluence increases. At higher fluences the relaxed forms have almost the same increase for both Ar-ion and N-ion cases. This is a clear indication of the ion direct interaction with DNA responsible for the DNA strand breaks. It is noticed that in our PIII of DNA the linear form of DNA appears negligibly whereas in our pervious experiment on ion beam bombardment of DNA using similar ion energy and fluence the linear form was observed in the electrophoresis [50]. The reason is thought to be that in PIII the ion energy has a distribution with a non-negligible low-energy component, thus the ion fluence with the ion peak energy is actually much lower than that calculated in operation, but in normal ion beam implantation there is no this problem. This fact further demonstrates the direct ion interaction with DNA dominating the DNA breaks. DNA breaks are potentials for mutation to occur. The DNA sequencing analysis revealed that some fragments of the DNA extracted from the bacterial mutant was different from those of the original DNA, indicating misrepairs which are responsible for mutation.

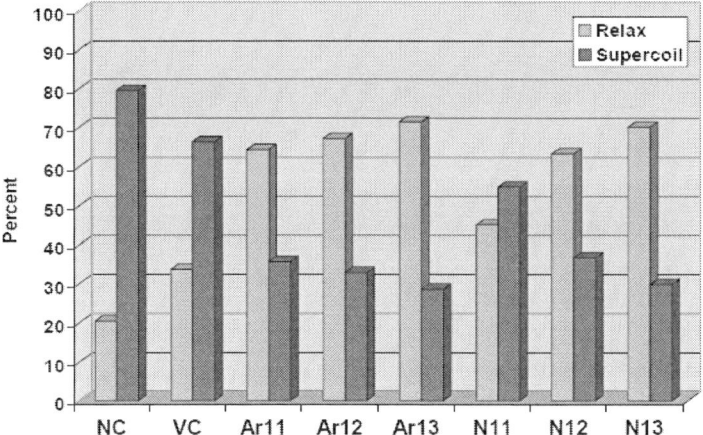

Figure 45. Graphic percentages of the DNA forms under different conditions: natural control, internal control (vacuum control) and after bombardment with ether nitrogen or argon ions to fluences of $10^{11}$, $10^{12}$ and $10^{13}$ ions/cm$^2$. NC: natural control, VC: vacuum control, Ar11: Ar-ion bombarded to $1 \times 10^{11}$ ions/cm$^2$, Ar12: Ar-ion bombarded to $1 \times 10^{12}$ ions/cm$^2$, Ar13: Ar-ion bombarded to $1 \times 10^{13}$ ions/cm$^2$, N11: N-ion bombarded to $1 \times 10^{11}$ ions/cm$^2$, N12: N-ion bombardde to $1 \times 10^{12}$ ions/cm$^2$, N13: N- ion bombarded to $1 \times 10^{13}$ ions/cm$^2$.

### 4.7.4. Conclusion

The effect of very-low-energy and low-fluence ion bombardment of naked DNA was investigated for exploring mechanisms involved in low-energy ion beam induced mutation. It is concluded that low-energy low-fluence ions which interact with DNA in the nanometer scale can indeed produce DNA damage in the forms of SSB, DSB and multiple DSB, which are the bases of mutation of biological organisms. The study confirms that one of the physical mechanisms in ion beam mutation is a small portion of incident ions capable of penetrating the materials covering the nucleus to directly interact with DNA and thus cause mutation.

Naked DNA is not natural and thus effects of ion interaction with the environment of DNA are ruled out. Therefore, next investigation is using very-low-energy ions to bombard DNA inside the living cells.

## 4.8. MOLECULAR DYNAMICS SIMULATION

Since at present it is technically difficult in determining the intrinsic features of ion-induced DNA changes, computer simulation becomes a very useful tool to assist in finding answers. Following the achieved research progress in experiments, molecular dynamics simulation (MDS) of very-low-energy ion interaction with DNA has also been involved in the research [57,58].

### 4.8.1 Methodology

To simulate ion bombardment of DNA in vacuum, DNA in A form that is the DNA form in low humidity or low pressure environment should be constructed. For investigating effect on the nitrogenous bases, 20 base pairs of alternating poly-AT and poly-GC double strands were constructed using HyperChem 7.0. MDS was performed starting with pre-equilibrium heating for 200 ps from 0 to 323 K (approximate temperature in the bombarding chambers), following by equilibration until the equilibria longer than 500 ps were achieved with time steps of 0.001 ps. The energy minimizations and MDS were performed *in vacuo* to imitate the dried and evacuated condition in experimental bombarding chambers. Both minimizations and MDS were done in AMBER 9 software package [59]. For investigating effect on various bonds, a 30-base-pair-long DNA duplex was constructed in A-form. The selected part is the residues number 760 − 789 of the green fluorescent protein plasmid (pGFP) in the GenBank, sequenced by Chalfie et al. [60]. The DNA duplex was built in Discovery Studio 1.7 software package [61]. The CHARMm27 force field [62] was applied on this molecule. To obtain the DNA structure in the equilibrium state in vacuum, the energy minimization, heating, equilibration and production in MDS were performed using Standard Dynamic Cascade protocol. The steps of energy minimization were divided into two parts: 1,000 steps of steepest descent minimization, and 4,000 steps of adopted bases Newton-Raphson minimization. Afterwards, the heating was performed for 60 ps from 0.0 to 323.0 K according to the experimental temperature. Then, the equilibration was performed for 2,900 ps at 323.0 K. And finally, the production was performed for 40 ps at the same constant temperature.

The velocity, $v$, of an ion was determined from the beam energy by $v = (2E/m)^{1/2}$, where $E$ is the kinetic energy of the ion, $m$ is the mass of the ion. Two sets of ion parameters were used: carbon ion with energy of 2, 20, and 200 eV and nitrogen ion with 0.1, 1, 10 and 100 eV, the former for bombarding the bases and the latter for bombarding the various bonds of DNA. The ion velocity direction was set to be random at an arbitrary atom of DNA. MDS was performed using combined quantum mechanics and molecular mechanics (QM/MM) coupled potentials. The QM region included only the ion. The energy and geometry of the region were calculated by the PM3 semi-empirical Hamiltonians. The long range QM-QM and QM-MM electrostatic interactions were calculated by Ewald sum. Figure 46 shows the program-constructed DNA structure and ion incidence.

## 4.8.2. Results and Discussion

### *4.8.2.1. C-Ion Irradiation Effect on Nitrogenous Bases*

The root mean square displacements (RMSD) of the backbone atoms of poly-AT were found remaining in small fluctuation after 1.0 ns of the equilibration, while those of poly-GC were stable after 1.5 ns of the equilibration, as shown in Figure 47. This indicates the poly-AT DNA double strands more inertia than the poly-GC backbones.

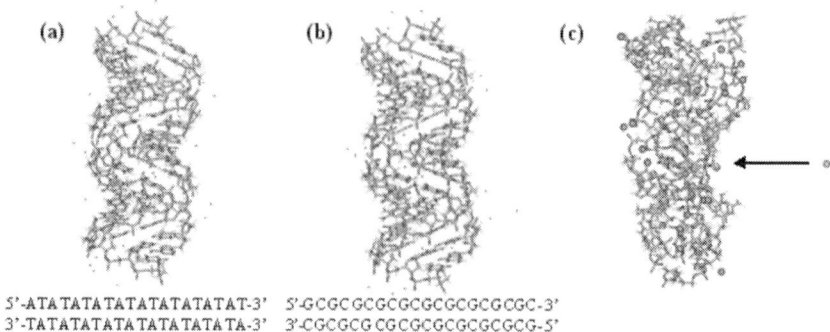

5'-ATATATATATATATATATATAT-3'   5'-GCGCGCGCGCGCGCGCGCGC-3'
3'-TATATATATATATATATATA-3'    3'-CGCGCGCGCGCGCGCGCGCG-5'

Figure 46. Construction of DNA and ion. (a) Poly-AT and (b) poly-GC A-DNA double strands. (c) Incident direction of ion.

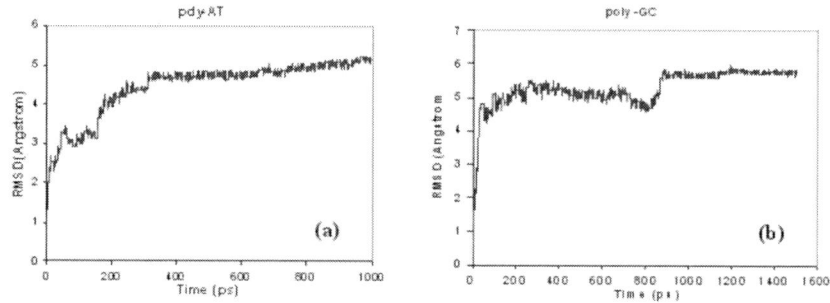

Figure 47. The RMSD of (a) poly-AT and (b) poly-GC DNA double strands.

**Table 5. Distance (angstrom, Å) between the backbone termini of two DNA strand after 150 ps MDS. In the case of 200-eV C-ion bombarding poly-GC, the ion passed through DNA after 150 ps simulation. The average distance change is the ratio of the difference between the mean distance of all non-zero energies and the distance of the zero energy over the latter**

| Distance between | Ion energy (eV) | | | | Average distance change (%) |
|---|---|---|---|---|---|
| | 0 | 2 | 20 | 200 | |
| Poly-AT A1-T40 | 11.7 | 12.6 | 13.0 | 11.1 | 4.56 |
| Poly-AT T20-A21 | 12.0 | 17.4 | 17.6 | 16.3 | 42.5 |
| Poly-GC G1-C40 | 16.0 | 19.1 | 18.1 | - | 16.3 |
| Poly-GC C20-G21 | 15.2 | 15.2 | 16.4 | - | 3.95 |

The tendency of DNA strand splitting was inspected by measuring the distance between the backbone termini of two strands, corresponding to A1-T40 and T20-A21 distances for poly-AT and G1-C40 and C20-G21 distances for poly-GC. The results are shown in Table 5. It is seen that the poly-AT's T20-A21 backbone termini is the most sensitive to the ion irradiation as it exhibits the largest distance increase subjected to C-ion bombardment.

The RMSDs of the base rings were measured to track the flexibility of bases. As seen from Figure 48, the behaviors of RMSD of poly-AT and poly-GC are different. The base rings of poly-GC are quickly stabilized after about only 5 ps of ion bombardment, whereas those of poly-AT take the time more than ten times as poly-GC takes to stabilize. The average RMSD of poly-AT is about one angstrom more than that of poly-CG. All of these results indicate that poly-AT is more unstable and more tend to be broken than poly-GC when subjected to ion attack.

### *4.8.2.2. N-Ion Interaction Preference to DNA Atoms*

Radial distribution functions (RDF), the distances of maximum RDF, $r_{max}$, and the RDF integrals are studied after Monte Carlo simulation of N-ion bombardment of DNA, as shown in Table 6. The results shown in the table are the mean values of two different doses. The higher RDF integral indicates the higher absorption preference of the implanted ion. It is seen that the preference of N-ion interaction with the DNA atoms is in an order of OP, O, O', N, C and C'. The shortest $r_{max}$ of OP also indicates the strongest interaction with the incident ion as the distance represents the distance between the atom and the ion, obviously, the shorter the stronger the interaction force.

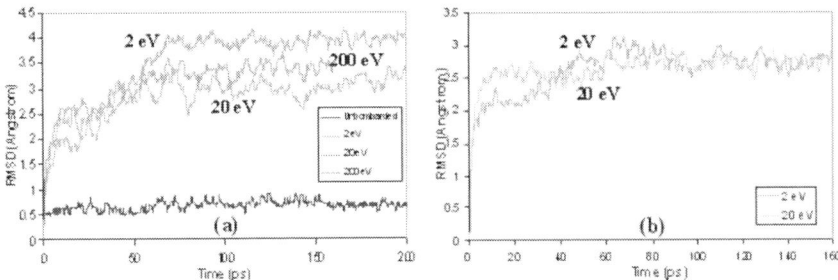

Figure 48. The RMSD of the non-hydrogen atoms in the base rings of (a) poly-AT and (b) poly-GC.

**Table 6. The mean values of the distance of maximum radial distribution functions, $r_{max}$, and the integral of radial distribution functions from 0.0 to 4.0 Å, $I_{4Å}$, of each atom type**

| Atom type | $r_{max}$ (Å) | $I_{4Å}$ (Å) |
|---|---|---|
| N | 3.85 | 1.83 |
| O | 4.75 | 2.8 |
| O' | 4.45 | 2.6 |
| OP | 3.3 | 3.25 |
| C | 3.9 | 1.5 |
| C' | 4.0 | 1.43 |

**Table 7. The bond lengths measured after certain time of N-ion irradiation**

| Bond type | Average equilibrium length (Å) | Modal bond length after ion irradiation (Å) | | | Mean increase (%) |
|---|---|---|---|---|---|
| | | 0.1 eV, after 10 ps | 1 eV, after 10 ps | 10 eV, after 6 ps | |
| O-P | 1.582 | 1.618 | 1.702 | 1.698 | 5.73 |
| O-P (ar) | 1.486 | 1.498 | 1.498 | 1.481 | 0.43 |
| C-C | 1.518 | 1.560 | 1.570 | 1.544 | 2.64 |
| C-N | 1.490 | 1.489 | 1.543 | 1.515 | 1.72 |
| C-O | 1.433 | 1.445 | 1.432 | 1.459 | 0.86 |
| C-C (ar) | 1.387 | 1.426 | 1.408 | 1.399 | 1.73 |
| C-N (ar) | 1.351 | 1.388 | 1.381 | 1.336 | 1.28 |
| C=O | 1.230 | 1.220 | 1.221 | 1.218 | -0.84 |

## (III). N-Ion Irradiation Effect on DNA Bonds

The ranges and medians of bond lengths of eight types were studied. The studied types included oxygen-phosphorus single bonds (O-P), oxygen-phosphorus aromatic bonds (O-P (ar)), carbon-carbon single bonds (C-C), carbon-nitrogen single bond (C-N), carbon-oxygen single bonds (C-O), carbon-carbon aromatic bonds (C-C (ar)), carbon-nitrogen aromatic bonds (C-N (ar)) and carbon-oxygen double bonds (C=O). The maximum, minimum and modal bond lengths in each bombardment were measured. Table 7 summarizes the main results. It is clearly seen that the O-P bond is the weakest as it has the largest increase in the bond length after ion attack, and following the O-P bond are the C-C, C-C (aromatic) and C-N bonds, whereas the C=O bond is the strongest.

### 4.8.3. Conclusion

Our MDS results show that for the nitrogenous base pairs, poly-AT is more sensitive to argon ion bombardment than poly-GC; for the phosphate group, deoxyribose (sugar) and bases, nitrogen ions interact in the preference sequence with OP, O (in base), O' (in sugar), N, C (in base) and C' (in sugar) atoms and most easily break the O-P bond and followed by C-C, C-C (aromatic) and C-N bonds. The findings demonstrate that low-energy ion beam induced DNA structural modification is not a random but preferential effect.

## CONCLUSION

Low-energy ion beam bombardment can induce mutation of biological organisms. The mutation is of broad spectrum — major advantages over other mutation methods. These features are due to the feature of four factors of ion implantation — mass deposition, energy deposition, momentum exchange and charge exchange. Direct ion interaction with DNA is a major mechanism for the induction of mutation

*Chapter 5*

# SUMMARY

Low-energy ion beam bioengineering is a newly and rapidly developed highly-interdisciplinary field, an expansion of ion beam modification of solid materials to biological living materials. It has great potentials for physics to serve applications in mutation breeding, gene transfer, biology analysis, life science and medical therapy. At Chiang Mai University, Thailand, research and applications of ion beam bioengineering have been vigorously carried out, particularly in the areas of ion beam inductions of mutation and gene transfer and also related fundamentals. Many have been achieved and many unknowns on relevant mechanisms are still open to investigate.

## ACKNOWLEDGMENTS

We wish to thank Prof. Yu Zengliang of the Chinese Academy of Sciences and his colleagues for many concerns, discussions, supports and collaborations. We thank researchers, graduate and undergraduate students of Chiang Mai University and other institutions from both Thailand and abroad for involvement in the research programs, particularly memtioned A. Krasaechai, S. Sangyuenyongpipat, I.G. Brown, B. Phanchaisri, R. Chandet, V.S. Lee, P. Nimmanpipug, S. Singkarat, S. Mahadtanapuk, K. Prakrajang, R. Norarat, S. Sarapirom, S. Jamjod and C. Wicharatana. The work has been supported by Thailand Research Fund, National Research Council of Thailand, Thailand Center of Excellence in Physics, and International Atomic Energy Agency.

# REFERENCES

[1] Wang, X. D.; Wu, Y. J.; Yu, Z. L.; et al. *Anhui Agricultural Science* (in Chinese), 1988.
[2] Wu, Y. J.; Yu, Z. L.; et al. *Anhui Agricultural Science* (in Chinese), 1989.
[3] Yu, Z. L.; et al. *Anhui Agricultural Science* (in Chinese), 1990.
[4] Yu, Z. L.; Qiu, L. J. and Huo, Y. P. *J. of Anhui Agricultural College* 1991, *18*(4) 251-257.
[5] Tanaka, R.; et al.; *Proc. $12^{th}$ Intern. Conf. On Cyclotrons and Their Applications*, Berlin, Germany, 1989, p.566.
[6] Hei, T. K.; Wu L. J.; Liu, S. X.; et al. *Proc. Natl. Acad. Sci. USA*, 1997, *94* 3765–3770.
[7] Electronics and Physical Sciences Research Council (EPSRC), UK (2004). *UK Research Network on Bio-Medical Applications of MeV Ion Beams*, http://www.ee.surrey.ac.uk/IBC/index.php?target=6:63.
[8] Hofsass, H.; Ronning, C.; Moller, W.; Homeier, H.; Stritzker, B. and Lindner, J. (Eds.) *Physics with Ion Beams, Innovative Research in Germany*; Druckhaus Fromm, Osnabruck, Germany, 2003.
[9] Vilaithong, T. Final Report to Thailand Research Fund on Research Project of *Low Energy Ion Beam Bombardment of Biological Cell*; Bangkok, July 2003.
[10] Vilaithong, T. Final Report to Thailand Research Fund on the Research Project of *Low Energy Ion Beam in Biotechnology*; Bangkok, August 2006.
[11] Yu, Z. L., translation eds.: Yu, L. D.; Vilaithong, T.; Brown, I. *Introduction to Ion Beam Biotechnology* (English Edition); Springer Science & Business Media, New York, 2006.

[12] Yu, L. D.; Sangyuenyongpipat, S.; Sriprom, C.; Thongleurm, C.; Tengsirivattana, C.; Suwanksum, R. and Vilaithong, T. *Nucl. Instr. Meth.* B 2007, *257* 790-795.
[13] Sangyuenyongpipat, S.; Yu, L. D.; Vilaithong, T.; Brown, I. G. *Nucl. Instr. Meth.* B 2007, *257* 136-140.
[14] Sangyuenyongpipat, S.; Vilaithong, T.; Yu, L. D.; Yimnirun, R.; Singjai, P.; Brown, I. G. *Solid State Phenomena* 2005, *107* 47-50.
[15] Davydov, S.; Yu, L. D.; Yotsombat, B.; Intarasiri, S.; Thongleurm, C.; A-no, V.; Vilaithong, T.; Rhodes, M. W. *Surf. Coat. Technol.* 2000, *131* 558-562.
[16] Yu, L. D.; Phanchaisri, B.; Anuntalabhochai, S.; Apavatjrut, P.; Vilaithong, T.; Brown, I. G. *Surf. Coat. Technol.* 2002 *158-159* 146-150.
[17] Yu, L. D.; Sangyuenyongpipat, S.; Annuntalabhochai, S.; Phanchaisri, B.; Vilaithong, T. and Brown, I. G. *Surf. Coat. Technol.* 2007, *201* 8055-8061.
[18] Vilaithong, T.; Yu, L. D.; Alisi, C.; Phanchaisri, B.; Apavatjrut, P.; Anuntalabhochai, S. *Surf. Coat. Technol.* 2000, *128-129* 133-138.
[19] Apavatjrut, P.; Alisi, C.; Phanchaisri, B.; Yu, L. D.; Anuntalabhochai, S. and Vilaithong, T. *ScienceAsia* 2003, *29*(2) 99-107.
[20] Anuntalabhochai, S.; Chandej, R.; Phanchaisri, B.; Yu, L. D.; Vilaithong, T.; Brown, I. G. *Appl. Phys. Lett.* 2001, *78*(16) 2393-2395.
[21] Phanchaisri, B.; Yu, L. D.; Anuntalabhochai, S.; Chanej, R.; Apavatjrut, P.; Vilaithong, T.; Brown, I. G. *Surf. Coat. Technol.* 2002, *158-159* 624-629.
[22] Anuntalabhochai, S.; Chandej, R.; Sanguansermsri, M.; Ladpala, S.; Cutler, R. W. and Vilaithong, T. *Surf. Coat. Technol.* 2009, *203*(17-18) 2521-2524.
[23] Implant Sciences (2003). PROFILE Code Software, Version 3.60, Wakefield, MA 01880-1246, USA.
[24] Yu, L. D.; Vilaithong, T.; Phanchaisri, B.; Apavatjrut, P.; Anuntalabhochai, S.; Evans, P.; Brown, I. G. *Nucl. Instr. Meth.* B 2003, *206* 586-590.
[25] Vilaithong, T.; Yu, L. D.; Apavatjrut, P.; Phanchaisri, B.; Sangyuenyongpipat, S.; Anuntalabhochai, S.; Brown, I. G. *Radiation Phys. Chem.* 2004, *71/3-4* 927-935.
[26] Sangyuenyongpipat, S.; Yu, L. D.; Vilaithong, T.; Verdaguer, A.; Ratera, I.; Ogletree, D. F.; Monteiro, O. R. and Brown, I. G. *Nucl. Instr. Meth.* B 2005, *227* 289-298.

[27] Alberts, B.; Bray, D.; Lewis, J.; Raff, M.; Roberts, K.; Watson, J. D. *Molecular Biology of the Cell*; Garland Publishing, New York & London, 1992, pp 11137.
[28] Voet, D. and Voet, J. G. *Biochemistry*; John Wiley & Sons, New York, 1990, pp 254.
[29] Sangyuenyongpipat, S.; Yu, L. D.; Vilaithong, T. and Brown, I. G. *Nucl. Instr. Meth.* B 2006, *242*(1-2) 8-11.
[30] Sangyuenyongpipat, S.; Yu, L. D.; Brown, I. G.; Seprom, C.; Vilaithong, T. *Nucl. Instr. Meth.* B 2007, *257* 136–140.
[31] Prakrajang, K.; Wanichapichart, P.; Anuntalabhochai, S.; Yu, L. D. *Nucl. Instr. Meth.* B, 2009, 267 1645-1649.
[32] Wanichapichart, P. and Yu, L. D. *Surf. Coat. Technol.* 2007, *201* 8165-8169.
[33] Anuntalabhochai, S.; Chandej, R.; Phanchaisri, B.; Yu, L. D.; Promthep, S.; Jamjod, S. and Vilaithong, T. *Proceedings of the 9$^{th}$ Asia Pacific Physics Conference*, Hanoi, Vietnam, October 25-31, 2004, Session 10: Applied Physics, 10-24C.
[34] Phanchaisri, B.; Chandet, R.; Yu, L. D.; Vilaithong, T.; Jamjod, S.; Anuntalabhochai, S. *Surf. Coat. Technol.* 2007, *201* 8024-8028.
[35] Doyle J. J. and Doyle, J. L. *Phytochem. Bull.* 1990, *19* 11.
[36] Anuntalabhochai, S.; Chandej, R.; Chiangda, J.; Apavatjrut, P. *Acta Horticulturae* 2000, *575* 253.
[37] Sanger, F.; Nicklen, S.; Coulson, A. R. *Proc. Natl. Acad. Sci.*, USA 1977, *74* 5463.
[38] Krasaechai, A.; Yu, L. D.; Sirisawad, T.; Phornsawatchai, T.; Bundithya, W.; Taya, U.; Anuntalabhochai, S.; Vilaithong, T. *Surf. Coat. Technol.* 2009, *203* 2525-2530.
[39] Mahadtanapuk, S.; Sanguansermsri, M.; Yu, L. D.; Vilaithong, T. and Anuntalabhochai, S. *Surf. Coat. Technol.* 2009, *203* 2546-2549.
[40] Anuntalabhochai, S.; Sitthiphrom, S.; Thongtaksin, W.; Sanguansermsri, M.; Cutler, R. W. *Scientia Horticulturae* 2007, *111* 389–393.
[41] Dubbern de Souza, F. H. and Marcos-Filho, J. *Revta brasil. Bot., Sao Paulo* 2001, *24*(4) 365-375.
[42] Wei, Z. Q.; Xie, H. M.; Han, G. W.; Li, W. J. *Nucl. Inst. and Meth.* B 1995, *95* 371-378.
[43] Zhao, Y.; Tan, Z.; Qiu, G. Y.; Du, Y. H. *Nucl. Instr. Meth.* B 2003, *211* 211.
[44] Hunniford, C. A.; Timson, D. J.; Davies, R. J. H. and McCullough, R. W. *Phys. Med. Biol.* 2007, *52* 3729.

[45] Chen, Y.; Jiang, B.; Chen, Y.; Ding, X.; Liu, X.; Chen, C.; Guo, X.; Yin, G. *Radiat. Environ. Biophys.* 1998, *37* 101.
[46] Lacomb, S.; Le Sech, C. and Esaulov, V. A. *Phys. Med. Biol.* 2004, *49* N65.
[47] Deng, Z. W.; Bald, I.; Illenberger, E. and Huels, M. A. *Phys. Rev. Lett.* 2005, *95* 153201.
[48] Chacon, F. A. *Ion induced radiation damage on the molecular level*, Ph.D. Thesis, University of Groningen, 2007.
[49] Norarat, R.; Semsang, N.; Anuntalabhochai, S. and Yu, L. D. *Nucl. Instr. Meth.* B 2009, *267* 1650.
[50] Yu, L. D.; Kamwanna, T. and Brown, I. *Physics in Medicine and Biology* 2009, *54* 5009.
[51] Nastasi, M., Mayer, J. W. and Hirvonen, J. K. *Ion-Solid Interactions: Fundamentals and Applications*, Cambridge University Press, 1996.
[52] Ziegler, J. F.; Ziegler, M. D. and Biersack, J. P. *The stopping and ranges of ions in matter*, *SRIM-2006*, 2006; see also www.srim.org.
[53] Biersack, J. P. *Z. Phys.* A 1982, *305* 95.
[54] Boege, F.; Straub, T.; Kehr, A.; Boesenberg, C.; Christiansen, K.; Andersen, A.; Akob, F. and Kohrle, J. *J. Biol. Chem.* 1996, *271* 2262.
[55] Bailly, C. *Methods Enzymol.* 2001, *340* 610.
[56] Mandelkern, M.; Elias, J.; Eden, D.; Crothers, D. *J. Mol. Biol.* 1981, *152* (1) 153. PMID 7338906 (http://www.ncbi.nlm.nih.gov/pubmed/ 7338906), http://en.wikipedia.org/ wiki/DNA.
[57] Ngaojampa, C.; Nimmanpipug, P.; Yu, L. D.; Anuntalabhochai, S.; Lee, V. S. *J. of Molecular Graphics and Modelling* 2010, *28* 533–539.
[58] Ngaojampa, C.; Nimmanpipug, P.; Yu, L. D.; Anuntalabhochai, S.; Lee, V. S. *Nucl. Instr. Meth.* B 2011, in press.
[59] Case, D. A.; Darden, T. A.; Cheatham, T. E. III; Simmerling, L. C.; Wang, J.; Duke, R. E.; Luo, R.; Merz, K. M.; Wang, B.; Pearlman, D. A.; Crowley, M.; Brozell, S.; Tsui, V.; Gohlke, H.; Mongan, J.; Hornak, V.; Cui, G.; Beroza, P.; Schafmeister, C.; Caldwell, J. W.; Ross, W. S.; Kollman, P. A. *Amber 8,* University of California, San Francisco, 2004.
[60] Chalfie, M.; Tu, Y.; Euskirchen, G.; Ward, W. W.; Prasher, D. C. *Science* 1994, *263* 802-805.
[61] Accelrys, *Discovery Studio Release Notes*, Release 2.0, Accelrys Software Inc.: San Diego, 2007.
[62] Foloppe, N.; MacKerell, A. D. *J. Comp. Chem.* 2000, *21* 86-104.

# INDEX

## A

absorption, 63
accelerating voltage, 8, 10
acceleration, 44
accelerator, 1, 52
acid, 16, 18, 38
activation, 16
advantages, 29, 65
AFM, 8, 9, 10, 22, 24, 26
Africa, 3
agricultural, 1
agricultural crop, 1
agriculture, vii, 1, 4
AMBER, 60
amino, 38, 57
amino acid, 38, 57
amino acids, 57
ampicillin resistance, 16, 18
amylopectin, 35
annealing, 35, 38, 46
antagonist, 75
antagonistic, 46, 47
antagonists, 46
anthocyanin, 38
anthracnose, 75
antibiotic, 18
antifungal activity, 47, 48
application, 4, 46
argon, 24, 53, 55, 58, 59, 65

Asia, 73
Asian, 2
atmosphere, 5, 8
atomic force, 8, 22, 24
atomic force microscope, 8
atomic force microscopy (AFM), 22, 24
atoms, 19, 23, 50, 61, 63, 65
Australia, 3

## B

*B. licheniformis*, 46, 47
*Bacillus (B.) licheniformis*, 46
backbone, 61, 62, 63
backscattering, 21
bacteria, 2, 4, 16, 17, 18, 46, 47, 58
bacterial, 14, 16, 22, 46, 58, 59
bacterial cells, 14, 16, 22
bacterium, 47
base pair, 43, 52, 60, 65
beam line, vii, 2, 6, 8, 9, 10, 14, 52, 53
beams, 1, 8, 25, 29, 39
behavior, 24
behaviors, 63
bending, 8, 10
bias, 53, 58
biocompatibility, 24
bioengineering, vii, 2, 9, 10, 14, 17, 52, 53, 67
biological control, 46

# Index

biological effect, 3
biological organism, vii, 1, 2, 3, 6, 31, 60, 65
biological systems, 38
biology, vii, 1, 55, 67
biosynthesis, 38
biotechnology, 1, 8, 38
bombardment, vii, 1, 5, 10, 13, 14, 15, 17, 19, 22, 23, 24, 25, 26, 28, 29, 31, 32, 35, 36, 37, 39, 43, 44, 45, 46, 50, 51, 53, 54, 55, 56, 57, 59, 60, 63, 64, 65
bond, 64, 65
bonds, 19, 35, 60, 61, 64, 65
Brazil, 3
breeding, vii, 1, 3, 67
broad spectrum, 65
bud, 31
buffer, 54

## C

capacitance, 24, 25, 27, 29
carbon, 53, 61, 64
CEA, 27
cell, vii, 2, 3, 6, 13, 15, 16, 19, 20, 21, 22, 23, 24, 29, 47, 49, 50
cell envelope, vii, 2, 6, 13, 15, 19, 20, 21, 22, 23, 24, 29, 50, 75
cell wall, 3, 19, 20, 22, 24, 49, 75
cellulose, 19, 20, 23, 24, 25, 26, 27, 28
chamber, 7, 8, 9, 10, 24, 32, 35, 53, 58
channels, 50
charged particle, 2
CHARMm27 force field, 61
Chiang Mai University (CMU), 2, 75
China, 1, 2, 3, 4
Chinese Academy of Sciences, 1, 7, 69
chitosan, 24, 27
chrysanthemum, 39, 41
cloning, 45
codes, 52
*Colletotrichum musae*, 47, 48
*Colletotrichum* sp., 46
*Columbia*, 3
Columbia University, 3

community, 4
components, 35
composition, 19, 23, 24, 51
computer simulation, 52, 60
concentration, 21, 25, 28, 52
conductance, 24
conflict, 55
contact angle, 24, 25
contaminant, 25
contamination, 6
control, 7, 10, 14, 15, 25, 26, 34, 36, 37, 40, 41, 42, 43, 44, 45, 46, 47, 53, 54, 56, 57, 59
control group, 43, 44
conversion, 35
cooling, 10
copper, 10, 11, 32
corn, 11, 14, 15, 16
cost, vii
covering, 60
crack, 50
crops, 1, 4
crossing pollination, 75
cucumber, 43, 44
culm, 36
cultivation, 35
culture, 46, 54, 58
culture media, 54, 58
*Curcuma*, 14, 20
*Curcumis sativus*, 43
current, 6, 8, 10, 32
cytochrome, 38
cytochrome P450, 38, 75

## D

damping, 10
database, 38, 46
death, 15, 16
deformation, 14
density, 8, 19, 49, 51
deoxyribose, 65
deposition, 50, 54, 65
developing countries, 4
*Discovery*, 61, 74

# Index

Discovery Studio, 61, 74
diseases, 46
distribution, 8, 20, 21, 49, 59, 63, 64
distribution function, 63, 64
disturbances, 10
DNA, v, vii, 4, 13, 14, 16, 17, 18, 23, 24, 25, 28, 29, 34, 35, 37, 38, 42, 46, 50, 51, 52, 53, 54, 56, 57, 58, 59, 60, 61, 62, 63, 64, 65, 74
double bonds, 64
double strand break (DSB), 54
Doyle and Doyle method, 38
duoplasmatron, 10
duration, 14

## E

E. coli, 14, 16, 17, 18, 22, 38, 52, 54, 57, 58
earth, 1
economic development, 2
Eden, 74
electrical properties, 27
electron, 1, 20, 22, 50
electron microscopy, 20, 22
electrons, 50
electrophoresis, 16, 17, 25, 28, 38, 54, 56, 58
electroporation, 29, 38
electrostatic interactions, 61
embryo, 14, 15, 16, 20, 31, 35, 44, 49, 50
embryos, 15, 31, 32
emission, 50
encoding, 46
endosperm, 38
energy, vii, 1, 6, 8, 13, 16, 18, 19, 21, 22, 23, 24, 29, 31, 32, 33, 35, 36, 43, 44, 45, 46, 47, 49, 50, 51, 52, 54, 57, 58, 59, 60, 61, 62, 65, 67
energy deposition, 50, 65
England, 2, 4
environment, 6, 8, 31, 60
enzymes, 17
equilibrium, 60, 64
equilibrium state, 61
equipment, 7, 10

Escherichia *coli* (*E. coli*), 14, 52
etching, 1, 3
Europe, 2, 4
evaporation, 5, 23, 31
exogenous, 6, 13, 14, 16, 17, 19, 23, 29
exposure, 11, 14, 58
extraction, 38

## F

facility, 7, 8, 10
Faraday cup, 8
February, 3
Fermi, 52
fertilizer, 47
film, 39
flavonoid, 38
flavonoid 3'hydroxylase, 38
flexibility, 63
flow, 53
flower, 38, 39, 40, 41, 42, 44, 46, 50
fluence, 6, 13, 15, 16, 21, 23, 24, 25, 26, 29, 32, 35, 43, 44, 46, 49, 50, 51, 54, 57, 59, 60
fluorescence, 57, 58
fragments, 54, 59
free radical, 50
free radicals, 50
fungi, 47, 48
fungus, 46
*Fusarium*, 47, 48
*Fusarium oxyaporum*, 47

## G

gas, 8
gel, 16, 18, 25, 54, 56, 58
gel electrophoresis, 16, 25, 54, 58
gels, 54
GenBank, 38, 61
gene, vii, 1, 2, 3, 4, 6, 16, 17, 18, 24, 25, 29, 45, 46, 47, 54, 55, 57, 67
gene cloning, 45
gene expression, 17, 54

gene transfer, vii, 1, 2, 3, 4, 6, 17, 24, 25, 29, 67
generation, 32, 33, 34, 35, 36, 39
genes, 16, 17, 18, 38, 45
genome, 35
genomic, 34, 35, 37, 46
gerbera, 39, 40
Germany, 2, 3, 71
germination, 15, 32, 43, 44, 46
GFP, 16, 17, 18, 55, 57
glass, 52
glucose, 17, 35
grain, 37
green fluorescent protein, 52, 58, 61
grounding, 58
groups, 7, 33, 52
growth, 18, 35, 43, 44
growth rate, 43, 44

## H

halogen, 10
harvesting, 36
HAT RAPD, 35
heat, 50
heating, 10, 52, 60
height, 7, 8, 36, 43, 44
helix, 35
horticulture, vii, 4
hot spring, 46
humidity, 60
hybrid, 44
hydrogen, 63
hydrogen atoms, 63
hydrophilicity, 24, 25, 29
hydrophobicity, 25, 29
HyperChem, 60
hypothesis, 55

## I

ice, 33
identification, 42, 46
identity, 38

image, 22, 26
images, 14, 20
immersion, 52, 53
impacts, vii
impedance, 24, 25, 27, 29
impedance spectroscopy, 24
in vitro, 25, 28, 46, 51
incidence, 32, 61
incubation, 16, 54
India, 4
indication, 59
indirect effect, 50
induce, vii, 1, 6, 13, 19, 29, 31, 45, 49, 51, 55, 58, 65
induction, vii, 2, 3, 14, 24, 29, 31, 32, 45, 57, 65, 67
industrial, 10
industrial application, 10
inert, 51
inertia, 61
inhibition, 47
initiation, 1
institutions, 69
interaction, vii, 2, 5, 22, 24, 50, 51, 55, 59, 60, 63, 65
interactions, 2, 61
interdisciplinary, 1, 2, 67
internal controls, 54
International Atomic Energy Agency, 4, 69
interval, 35
intrinsic, 60
iodine, 34, 35
Iodine, 35
ion beam biotechnology or bioengineering (IBB), 2
ion bombardment, 15, 17, 22, 23, 24, 25, 26, 28, 31, 35, 36, 37, 39, 43, 44, 46, 50, 51, 55, 56, 57, 59, 60, 63, 65
ion implantation, 1, 7, 10, 21, 49, 51, 52, 53, 65
ion implanter, 6, 7, 10, 11, 32, 35
ion source, 7, 8, 10
ion species, 8, 10, 13, 16, 22, 51, 52, 55
ions, vii, 2, 5, 10, 13, 14, 15, 16, 18, 19, 20, 21, 23, 24, 25, 26, 28, 31, 32, 34, 35, 36,

39, 43, 46, 49, 50, 52, 53, 55, 56, 57, 58, 59, 60, 65, 74
irradiation, 5, 31, 63, 64

## J

Japan, 2, 3
jumping, 53

## K

KDML 105, 32, 35, 36, 37, 38, 39
kilo electron volt (keV), 75
kilo volt (kV), 75
kinetic energy, 61

## L

*lacZ*, 16, 17, 57
lamina, 53
laminar, 53
linear, 19, 20, 54, 56, 59
lipase, 46, 47
*lipoic acid synthetase*, 16, 18
liquid nitrogen, 38
loading, 54
location, 5, 6, 31
London, 73
low-energy, vii, 1, 8, 19, 29, 31, 46, 47, 49, 50, 51, 52, 54, 57, 58, 59, 60, 65
Luo, 74
Luria broth (LB), 17

## M

macromolecules, 13, 14, 19, 29
magnet, 8, 10
magnetic, 8
magnets, 8
majority, 32
Mammalian, 4
manipulation, 10
manufacturer, 52

marker genes, 16, 18
markers, 17
materials, 1, 3, 5, 8, 14, 19, 25, 31, 49, 50, 53, 58, 60, 67
matrix, 19
measurement, 10, 17
measures, 31
meat, 43, 44
mechanism, 24, 65
media, 17, 54, 58
membrane, 24, 25, 26, 27, 28, 49
membranes, 24, 25, 28
micro/nano-craters, 22
microbes, 1, 7, 48
microinjection, 29
microphotographs, 50
microscope, 8, 10
microscopy, 20, 22, 24
microstructure, 23
modification, 1, 8, 14, 24, 35, 57, 65, 67
molecular dynamics, 60
molecular dynamics simulation, 60
molecular mass, 18
molecular weight, 37
molecules, 15, 19, 20, 22, 25, 35, 50, 53
momentum, 65
monolayer, 24
Monte Carlo, 63
morphology, 24
mung bean, 43, 44
mutant, 36, 37, 38, 39, 57, 59
mutants, 36, 38, 42, 57
mutation, vii, 1, 2, 3, 6, 10, 31, 32, 38, 45, 46, 49, 50, 51, 55, 57, 58, 59, 60, 65, 67
mutations, 6, 36, 46

## N

naked DNA, 51, 60
nanometer, 52, 60
nanometer scale, 60
nanometers, 19
NASA, 3
National Research Council, 69
natural, 42, 46, 54, 59, 60

NC, 59
New York, 4, 71, 73
Newton, 61
Newton-Raphson, 61
nitrogen, 10, 16, 18, 24, 28, 32, 34, 35, 36, 38, 39, 43, 44, 51, 53, 55, 58, 59, 61, 64, 65
normal, 5, 8, 10, 14, 19, 31, 35, 44, 57, 59
nuclear, 51
nuclei, 15
nucleotide sequence, 38
nucleus, 51, 60

## O

*O. sativa japonica*, 38
observations, 45
onion, 14, 21, 22
organ, 43, 44
organism, 3, 6
Oryza sativa indica, 75
ovaries, 44
ovary, 43, 44
oxygen, 64

## P

Pacific, 73
parallel, 20
particles, 1, 8, 23, 31
pathways, 6, 13, 19, 23, 29
PCR, 25, 28, 37, 38
penetration, 13, 19
Petri dish, 10
petunia, 40, 41, 42, 50
pGEM2, 16, 18
pGEM-T, 16, 17, 38, 46, 47
pGFP, 16, 17, 18, 25, 28, 52, 54, 57, 58, 61
phenotypes, 32
phenotypic, 36, 37, 38
phosphate, 65
phosphorus, 64
photoperiod, 35
photoperiod-insensitive, 35

physical mechanisms, 60
physicists, vii, 2, 5
physics, vii, 1, 5, 55, 67
pigments, 38
planar, 35
plant, 3, 7, 19, 20, 22, 24, 31, 33, 35, 36, 38, 43, 46, 47, 48, 49, 50
plasma, 52, 53, 58
Plasma and Beam Physics Research Facility, ix, 2
plasma immersion ion implantation implantation (PIII), 52
plasmid, 16, 17, 18, 25, 28, 38, 51, 52, 53, 54, 56, 57, 58, 61
plastic, 35, 39
play, 38, 50
PM3, 61
pollination, 43
polyacrylamide, 16
poly-AT, GC, 60, 61, 62, 63, 65
polymer, 35
polymorphisms, 43
porous, 5, 49, 50
powder, 38
power, 6, 49, 53, 58
PRAL, 52
preference, 63, 65
pressure, 6, 8, 14, 32, 53, 58, 60
primer, 34, 42, 46, 47
probability, 58
probe, 2
production, 61
PROFILE, 19, 21, 49, 52, 72
program, 50, 61
promoter, 57
proposition, 3
protein, 18, 47, 48, 57
protocol, 52, 61
pseudo, 51
pulse, 11, 35, 53
pumping, 6, 8, 39
purity, 54, 57
purple glutinous rice (Kum Doi Sa Ket), 32
pYES2, 47, 48
pYGFP, 16

pYlip, 16, 18
*Pyricularia grisea*, 47, 48

## Q

quantum, 61
quantum mechanics, 61
quantum and molecular mechanics (QM/MM), 61

## R

radial distribution, 64
radiation, 2, 3, 6, 13, 19, 72, 74
radiation damage, 6, 13, 19, 74
radicals, 50
radiofrequency (RF), 53
radiological, 3
radius, 52
random, 6, 20, 31, 35, 38, 46, 61, 65
random amplified polymorphic DNA, 35, 38, 46
range, 13, 19, 23, 29, 43, 46, 49, 50, 51, 52, 61
RAPD, 34, 35, 37, 38, 42, 46
reality, 19, 49
recommendations, iv
recovery, 6
regular, 24
regulation, 38
relaxed, 54, 58
reparation, 29
residues, 61
resistance, 16, 18
restriction enzyme, 17, 54
rice, 1, 7, 10, 11, 31, 32, 33, 34, 35, 36, 37, 38, 39, 50
rings, 63
risk, 3, 8
room temperature, 32, 47
root mean square displacements (RMSD), 61
rose, 39, 40
roughness, 24, 25, 26, 29

Rutherford, 21
Rutherford backscattering spectrometry (RBS), 21

## S

*Saccharomyces cerevisiae*, 4, 16
sample, 7, 8, 10, 11, 29, 32, 35, 52, 53, 54, 58
sample holder, 8, 10, 11, 32, 35, 52, 53, 58
scanning electron microscopy (SEM), 22
school, 2, 3
screening, 36, 40, 52
SDS, 16, 18
secondary effect, 5, 50
secondary electron, 50
seed, 10, 15, 31, 32, 33, 34, 35, 43, 44, 49, 50
seedlings, 32, 33, 35, 45
seeds, 1, 7, 10, 11, 15, 31, 32, 34, 35, 36, 38, 39, 43, 44
segregation, 33
SEM, 14, 22, 50
separation, 53
sequencing, 57, 59
shape, 6, 8, 10, 32, 40, 41
shoot, 45
simplified mean-pseudo-atom model, 51
simulation, 22, 23, 52, 60, 62, 63
single-strand break (SSB), 54
skin, 14, 22
sodium, 16
sodium dodecyl sulfate polyacrylamide gel electrophoresis (SDS PAGE), 16
software, 60
soil, 32, 35, 43, 44
South Africa, 3
species, 8, 10, 13, 14, 16, 22, 43, 51, 52, 55
spectroscopy, 24
spectrum, 46, 65
sputtering, 13, 19, 25, 53
SRIM, 52, 74
SSB, 54, 60
stabilize, 63
stages, 35

stainless steel, 53
Standard Dynamic Cascade, 61
standards, 14
starch, 35
steel, 8, 53
sterile, 7, 47, 52
stopping, 19, 49, 51, 74
strain, 14, 16, 18
structure, 20, 22, 23, 24, 35, 50, 58, 61
structuring, 23
students, 2, 69
sucrose, 54
sugar, 65
sulfate, 16
supercoiled, 55, 58
surface area, 25
surface roughness, 24, 25, 26, 29
survival, 14, 32, 43, 44
symptom, 46
synthesis, 38

## T

targets, 31
taste, 32
TEM, 20
temperature, 10, 14, 32, 35, 38, 46, 47, 60
testing, 27
Thai, 3, 4, 32, 34, 35, 36, 37
Thailand, vii, ix, 2, 4, 32, 67, 69, 71
therapy, 2, 3, 67
thermal energy, 58
Thomas-Fermi (TF) screening radius, 52
time periods, 58
tissue, 31, 38
traditions, 2
traits, 43, 44
transfer, vii, 1, 2, 6, 13, 14, 16, 17, 18, 19, 23, 24, 25, 28, 29, 54, 57, 67
transformation, 52
translation, 10, 71
transmission electron microscopy (TEM), 75
transportation, 32
tumor, 2

## U

UK, 71
ultraviolet (UV), 75
undergraduate, 69
University of Surrey, 3
USA, 2, 4, 43, 71, 72, 73
UV, 16, 54, 57, 58

## V

vacuum, 1, 5, 8, 14, 15, 31, 32, 37, 39, 49, 53, 54, 55, 58, 59, 60
values, 24, 63, 64
variant, 46
variation, 37, 38, 40, 41, 42, 45, 49
variations, 35, 36, 37, 38, 40, 41, 45, 46
vector, 38, 46, 47
vegetable, 43
velocity, 23, 61
vibration, 10
Vietnam, 73
*Vigna radiata*, 43, 44
voltage, 8, 10, 35, 54

## W

water, 5, 10, 14, 23, 25, 28, 31, 35, 39, 52, 54
water evaporation, 23
wild type, 32, 34, 43, 46

## X

X-ray, 50

## Y

yeast, 2, 4, 16, 18, 47
yield, 19
YPD, 17
Yu Zengliang, 1, 3, 4, 69